belly

belly

nora bateson

tp

Triarchy Press

Published In This First Edition In 2026
By
Triarchy Press
Axminster, England

tp

info@triarchypress.net
www.triarchypress.net

A catalogue record for this book is available from the British Library.

Cover Design & Production:
Noah Qwarfordt, Nora Bateson, Leslie Thulin Eubanks

Book Design & Production:
Leslie Thulin Eubanks

Editor:
Leslie Thulin Eubanks

Interior Artwork:
Noah Qwarfordt, Nora Bateson, Leslie Thulin Eubanks
Rachel Hentsch, Vivien Leung

Print ISBN: 978-1-917251-19-8
ePub ISBN: 978-1-917251-20-4

dedication

She fed me into life.

She fed me adventure.

She fed me provocations and honed my ability to fight.

She fed me bone, blood, and belly upset …

She fed me spinach with lemon butter.

She fed me the infinity of learning through caregiving when her body grew close to death.

She now feeds me her forever companionship from beyond this world.

This book is dedicated to my mother.
Lois Jane Cammack Bateson
(1929-2024)

table of contents

acknowledgments

This book began long before it began. And it began in many directions and in multiple contexts. This is in keeping with the movement of the book itself. I was asked by my kids to write a cookbook and simultaneously asked by my friend Yogi Hendlin to write a paper for the *Journal of Biosemiotics* on my work with Warm Data and nth-order agency. The two projects appeared utterly distinct. I did not notice their merging until it had already happened in my belly, then I noticed they had been fused in my earlier writings since 1993. This is a beautiful and strange perfection of overlap that has brought the book into the form of what my father, Gregory Bateson (1972) used to call a Metalogue (a metalogue is a wondrous parallel occurrence whereby the structure of a difficult conversation is relevant to the subject).

This book also began a long time ago when I made a cookbook with my friend Bancha Leelaguagoon. Together we made a CD-ROM non-linear ethnography that streamed together our friendship, our cultural confusions, and recipes for Thai food. My apartment smelled like garlic and chilies for 6 months.

This book began when I was in my mother's womb, receiving nutrients and beginning life. Then in my kitchen. I am grateful to the ancestors who brought fire and iron and water to root and fruit and herbs … My mother is everywhere in this book. I hope she is smiling somewhere.

I have a dear friend, who is also an angel of everything, Leslie Thulin Eubanks. We have had 1,000 long talks through this material, and 3,000 meals. Those around us think we work too hard, but they have no idea how much fun we are having. She gets me, she pushes me, she goes with me on all sorts of adventures, written and otherwise. Leslie has been with me through the editing of all of this, and we have spent hours, days, months, years even in discussion around bringing these ideas onto paper. She has been indefatigable in her dedication to tending to the technical and artistic detail of getting this book printed with all of its art, and design. My grammar is filled with invisible theoretical corrections, and Leslie is probably the only person who could possibly track these necessary phrasings I hold so passionately. I am grateful to her beyond words, and I know I am lucky to share my work with her.

Others who have helped edit the text include: Phillip Guddemi, Rex Weyler, Dani Rae Wilson, Rachel Hentsch, Sahra Bateson Brubeck, Trevor Brubeck, Howard Silverman, Peter Sforza.

I am also fortunate to work in the Warm Data team with some of my favorite artists. I knew when I invited Vivien Leung, Rachel Hentsch and Leslie Thulin Eubanks to help me illustrate this book, the years spent together with me in exploration of the theory and practice would guide the imagery. Or course, their art would shine with the non-verbal expressions of all that I was trying to say in words. And, I am especially excited to see the beautiful art that my stepson Noah Qwarfordt brought to this project. It is fitting that Belly has birthed a path for this talented artist to emerge.

Andrew Carey of Triarchy Press is my publisher, co-editor, and most trusted keeper of the delicate sinews of my heart as my words go into print. I will always remember the first time he read my pages for *Small Arcs of Larger Circles* and saw in me that which no other publisher did. Again, I cannot help but feel blessed. He has stood by me as I have broken with traditional book forms that limited me in the need for multiple forms of expression. We are both pleased that there is nothing in this book created with AI.

Through it all, and in and out of every day, my beloved husband Mats Qwarfordt has been there by my side. He keeps me real, brings me coffee, and even goes to his workshop late on the days I capture him in conversations when I need to think with him.

I am grateful for the
fruits, roots, beans, leaves, nuts, and berries
that bring food to us all.

a glass of water

Nobody would ever write a recipe for a glass of water, and yet perhaps that is exactly the angle from which to begin.

How do you make a glass of water? In the words of Carl Sagan, who talks about making an apple pie, "You must first invent a universe."

And for that matter, how do you *make* a change? How do you *make* things happen?

Our family gathers around meals, and I love to cook. So when my grown children asked me to write a cookbook for them, I set about finding out which of my recipes they wanted me to include. As the requests stacked up, so did the giggly jests about how I would write out the measurements when I never measure anything in the kitchen. My kids said things like, "She will probably write, 'just add some of this and a little of that.'"

They were not wrong.

I felt a thud in my belly at the thought of dissecting the amounts and timings of making these meals — something felt off.

While recipes are treasures, and I love to read them, they are not how to cook. The "how" of cooking lies elsewhere, and it is vitally wrapped up in a lifelong learning of how to be alive, how to be human, and how to perceive and respond to the beautiful and painful moments of a life.

The codification of a recipe is convenient. Without a practice of affection for the infinite details in the gesture of something as simple as making a glass of water, a recipe can be dangerous.

While that may sound like hyperbole, perhaps it is a small peek into the possibility of shifting some of the deep insidiousness of a world plagued with appetites for destruction, control, and manipulation. The rush to grab a quick fix and a one-size-fits-all how-to is a non-trivial dismissal of much essential learning.

Learning the rhythm of one's own body, the smell of the seasons, the weight of devotion — these are not written in the recipe. Do not mistake the recipe for the meal, the harvest, the ancestors, and the future.

Any artisan will tell you that their materials and their tools are not merely materials and tools — they are intimate relationships with ever-shifting circumstances that must remain able to be responsive. Some days the leather is cold, and the workshop must be warmed before it can stretch; some days the wood is dry, and the woodworker must use the right instrument for dry wood. Some tomatoes are sweeter, and more salt will be needed; every aspect of the project requires learning to sense contextual shifts.

Pouring a glass of water is both a movement that is habituated into a familiarity of thoughtlessness, or automated action, as well as an honoring of primal and sacred contact within the waters upon which all of life relies. The action of reaching for water is shared with the roots of plants and the long neck of the giraffe.

A glass, a vessel, a holding of that water is both as banal as any other item in the cupboard and as profound as the carrying of life from one generation into another through the mysteries of membranes that allow water to cross. It's just a glass of water. Why on earth would you need a recipe for that?

Perhaps there's another realm to attune to just behind — or beneath — the recipe. The way each action is steeped in detail speaks of who it is possible to be. There is a kind of violence in hacking at the respect held in the many stories of each "doing." And yet, in a culture addicted to ease and quick formulas, this violence becomes breakfast.

Facilitation, while appealing as a concept of making things easier, also disrupts the necessary depth of consideration across all relations, from the glass of water to the family, to sex, to world politics, and the entire idea of an economy. The faucet has made it easy to grab a glass of water. Hidden behind the shiny knobs lies the world of pipes and the remarkable feat of the development of plumbing.

The faucet has also removed relationships from the carrying of water, the noticing of the rushing of the river, the conversations with others gathering water, the weaving of the reeds, the making of gourds, and tending the path to the water source. The glass can be a beautiful and practical designed artifact, cupped to the size of a hand, made from the remarkable process of heating sand — and while the glass is in itself a thing of beauty, the cupping of hands that scoop down to bring water to your face, your lips, your hair … is also a remembrance of an ancient prayer to that which is central to life, water.

Be careful of ease, it can erase. The erasure is gushing with unthinkable destruction — a consequence, perhaps, of thoughtlessness. Utility that hacks the hard work and discomfort of thoroughness comes with consequences. Ease can also be the fruit of discipline and rigorous depth of practice.

Discernment is crucial. The difference between buying a plastic picnic cup and studying the art of ceramics for decades — knowing which clay to use, holding wet clay over the wheel, painting a cup, waiting for the kiln ... is a difference of saturated knowing. Not that there is anything wrong with plastic picnic cups, but that they are placeholders for another knowing, not replacements.

Tickling habituated assumptions may seem like a frivolous task in these times when the rush to get from one day to the next is in hyperspeed. And yet, there is exactly this need to unravel the disregard, banish the laziness, bring fascination where neglect took root, sensitize to the seductions of indifference, and hold, taste, breathe, and offer from the wild generosity of integrity. The toil is not a waste of time; it is contact with the effort needed to remember what matters. What matters? Every ancient culture has sayings about the recklessness of hurry.

This book became a recipe instead of a book of recipes — not to make the food on a plate, but of the grace of food, as it moves through generations, metabolisms, ideas of what an action is, an attunement to life. Each glass of water, each meal, each morsel that enters the mouth's universe of taste buds and travels the labyrinth of digestion is a portal to being alive within the gift of life. It is an offering of nothing less than a glass of water.

What is in the refrigerator is not just what is in the refrigerator, but an entire cosmos within which to discover being alive together in the softness of the belly.

I have no model for peace, only glimpses of it, metaphors for it, similes to what I cannot fully grasp and hold. Among them: a bowl of clear water. A boat drifting on a slow river. A lake among hills. The vast depths of the sea. A drop of water at the tip of a leaf. The sound of rain. The sound of a fountain. The bright dance of the water-spray from a garden hose, the scent of wet earth. (Le Guin, 2016, Blog post 119)

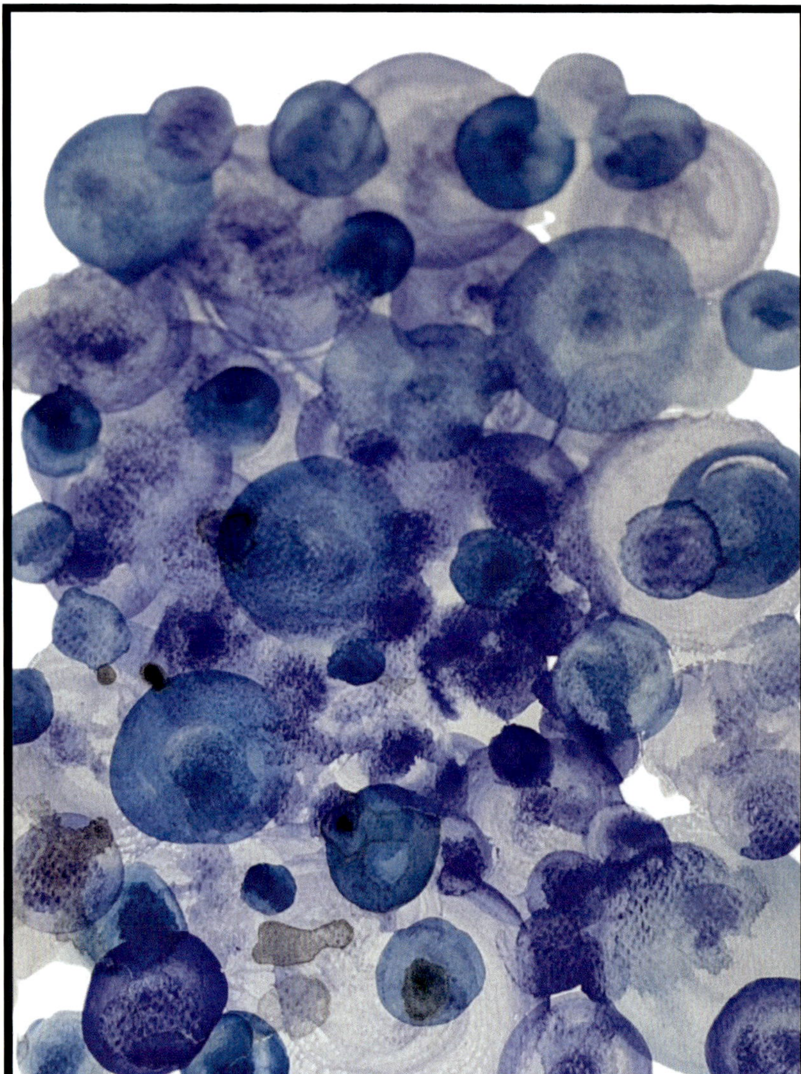

Fig. 1. Vivien Leung. (2025). [watercolor]

belly is the beast

Here in the middle of the mess.

Where all the bubbling takes place, this is where the breaking and the healing are brewing, stewing, and digesting.

This is the deep mischief.

go home

The morning's clinking of dishes wakes the ghosts of
yesterday.
Tiny threads of history wrapping around the arc of the day.

The traditions of tea, coffee, and breakfast in their details
hold the whole story — of why it is so hard to stop the
wars and horrors.

No point in making big graphs.
No point in huge committees.
Those big programs only enhanced the disaster.

Go home.

Touch the ghosts, find them, notice them hiding under
your sleeve and in your cupboards.

They are in the kitchen, feeding you daily doses of the past.
Comforting, familiar ideas of life where yesterday, the day
before, and tomorrow are all alike.
Whispering away the shake of shifting paths in illusions of
practicality and reason.

They say:
This is how it isssssssssssss.
No, it is not.

A visitor came ...
When asked what he would like for breakfast, he said,
"Nothing special — just cornflakes and chocolate milk."

Another visitor came ...
When asked what he would like for breakfast, he said,
"Nothing special — just buckwheat and pork."

After a while, all the visitors' breakfasts began to reveal
that each of us is living in a universe of our own, infected
perhaps by culture, and built in experience and language.

Which one is the real one?
Which one is right?
These are the trapping questions.

Humility. At breakfast.
Unfamiliar microthreads of possibility spread thinly over
toast.
A sip of warmed confusion.
Scrambled know-how with a side of fried habits.

Go home, to where the ghosts make breakfast.

Fig. 2. Vivien Leung. (2025). [watercolor]

threads

There is much to do, and the bulk of it is showing affection where it has been withheld.

These are the moments when it becomes clear what mattered.

Misplaced attention, misplaced concern, misplaced time spent gathering the wrong ingredients for tomorrow.

Destruction was left unbridled, greed has been substantiated.

And Rilke said, "Just give [us] a little more time."

One more chance.

Six more months and I know I can rearrange myself into a different way of being,

I will remember that while the state has been a mess, the spring still comes,

I will remember that the wars did not stop the eternal movement of the stars.

I can learn to tend my neighbor's pain. I can turn toward you and give as I have never given before.

rules

There are rules.

Then ... it is necessary to break the rules.

But there are rules for breaking the rules.

Because there are bigger rules —

and those rules cannot be contained by rules.

There are ways beyond rules that every practice has endeavored.

Because I am not complete, I am learning.

I want to offer you something that may bring you an experience of being within the grace of life.

This is what matters.

This is the rule.

Fig. 3. Vivien Leung. (2025). [watercolor]

my spine is louder than the teleprompter

Response to a situation is always limited by what possibilities are perceived.

There are rules for everything — formulas, models, and easy-to-use, step-by-step manuals promise successful endeavors in fitness, parenting, sustainable development, and financial security. Resonant through every corner of life is the message that, "If you do as the experts say, you will have success." In all my years as an educator of systems theory and complexity theory, I have found that this is something out of reach of analysis.

All the theory in the world does not seem to alter the anxiety of not having a plan. It is as though, physiologically, there has been an incorporation of the industrial need for linear strategy that at this point is presenting as somatic. The abstractions of the engineering blueprint have grafted into bone — the strategies are as sinew, the measurements as blood. I cannot simply learn some new techniques for finding solutions and be over and done with the last several hundred years of imbibing the logic of the factory.

Optimization, efficiency, time management, high yield, cost ratio — success is not just financial, it is identity … I do not know where I begin and where the contexts of modernity's hold on me ends. I have this mess in my hands, in my kitchen, in my marriage, in my body, in my belly.

My husband, Mats, who is an artisan, described for me the way his mentors in woodworking taught him his craft by sending him to many different master woodworkers, each of whom would approach a problem differently. In this way his learning included not only multiple choices of techniques to work with, but the larger learning that when one studies deeply, what is gained is not know-how but the knowing that each situation can be met in its own way by the craftsperson who will find a voice in the possibilities they see in the moment. This is the art form.

What is in the scope of viable next moves? The "next best thing" and the "relevant next" are not landscapes of all that is potentially doable, but rather landscapes of what is within the purview of the person or organism. When I hear the voicemail ask me to choose between dialing one for this, two for that, and three for the other ... I am from a generation that expected human beings to answer the phone. It seems so far away now. I still very often push zero for an agent.

I see the numbered options as obstructions to my task. My children do not. They have not lived in a world where there were actual people who would speak to them about their call (which undoubtedly does not fall under any of the prompt options). For them, what is possible is to choose one and then get shuffled around; for me, what is possible is to try to get to a real person as quickly as possible.

In the end, both of our approaches are probably equally frustrating and occasionally satisfying. I worry that the generations who have not lived through sorting out day-to-day issues with other human beings will not have the inclination to get angry at the digital apathy of AI and scream at corporate evasion of contact.

I remember a time when my daughter was about nine years old, and there was a new popular parenting book that many of her friends' parents were discussing in the parking lot. She came home one day and said, "Mom, I hate parenting books. Why does everyone read them?" She was angry because she felt the strangeness of people listening to a book and not her friends. How could the author of the book have known what was best for these actual people in actual situations that were in response to other situations? She was right. But what she could not know was the desperation that parents feel, trying to raise children with all the goodness they can, and not repeating the mistakes of the past.

There is a weird parenting competition among parents today, playing out various schools of thought through their kids, revealed through behavioral models. The kids are on display, and they are being measured, labeled, and assessed according to a selection of parenting plans. My daughter felt the judgment and the constriction — I cannot say she was wrong.

Response to a situation is always limited by what possibilities are perceived.

If I want to make a custard and I have never made one before, I am likely to find a recipe and try to get the basic idea of how a few people make custard, and then try it out. I think I am like most people in this approach. Most dishes that are not in my tried-and-true repertoire may require a little research.

Once having looked through a few versions, I might get some ideas of my own. Seeing that one recipe uses more eggs, or one uses coconut milk, I might find my own tastes leaning toward something that sounds good to me. I like vanilla — so I might add a little extra. I like the way salt and sugar create an intensity of flavor — I might be a little more liberal with adding those too.

But I will also notice the context I am making the custard for. Is it going into a pie, or will it be poured over a cake? Is it for little kids, or is it for a person with a sensitive palate? Although I may have preferences, the custard is likely going to be shared, and sensitivity to whom it is shared with is part of the reason for making it in the first place. Every aspect of making the custard is relational. The ratio between the eggs and milk, the sugar and salt, is part of a larger set of relationships between me and who I will be sharing it with, and that relationship is within another context of where our experiences with custard overlap, and don't overlap.

The timing is also within a group of relationships. If it is being served cool and thick, it will need time in the refrigerator; if it is being poured warm, it is better freshly made.

Parenting and custard are very alike in that, as I approach a how-to moment, I may look around to see what others have done. This is a needed and natural response to having been living in a society of isolation and individual failure and success. I did not grow up in a way that gave me the depth of confidence from generations of care, to know when what I am doing in the name of care is actually healthy care, and when what I am doing in the name of care is clutchingly overbearing, or capriciously negligent.

All the generations who came before us approached parenting the way the social pressures pushed them; they also did not have the innate perception to see the complexity of their children and tend to them as needed. Parents have had to prepare their children for the hard knocks of life, whether in ancient times or modern times. To get us through school and into jobs as "individuals," they measured us against each other, labeled us, and set the course of our lives by doing so.

For some, that was a boost into a social and financial higher club of pressures and one-upping; for others, it was condemnation to a life of humiliation and struggle. Neither necessarily became happy adults. How then should we know how to parent when the parent who came before entrained the logic of capitalistic, individualistic success into us?

Of course, there will be recipes. Of course, there will be manuals. The Industrial Revolution in the 1700s set that in motion. Everything became like a factory — everything became a product to bring to completion and set out into the world for harsh judgment. With an end product in mind, a linear strategy is required to achieve it. Whether the end product is a successful next generation or a dessert that the guests coo over, the ingrained, entrained, implicit learning is that a step-by-step process is needed to map out the best way to achieve it.

I keep thinking of my daughter. She felt so afraid of those parents who were stuck on the parenting books, afraid she would not be seen, afraid the situation in the world would be misunderstood by the abstraction in the book. The vulnerability of the contextual information getting fitted into the description in the manual when it does not fit ... dangerously delimits the scope of possible nexts.

The incredible trust with which children look to their parents evokes in me both an acute feeling of honor and an unflinching desire to never let them down. When my daughter looked up at me, what she wanted was for me to know in my belly how to be a good parent and be able to meet each moment with warmth and wisdom.

To keep her safe, I would need to not look through a frame of reference in a book, but look, really look at what was happening right there in the relationships. She was asking me to know how to help her find her way in the world with the same ancient knowing my body knew how to grow her inside.

But I did not have a path to that source of mothering, at first. It required practice — like my husband and the woodworking — it required seeing lots of parents do it differently, it required knowing my own voice, my own wounds, my confusions. My salt, my sugar.

It has taken years of scrutinizing, not my children, but what I have come to call the "cultural teleprompter" that is constantly putting words in my mouth that lead, at best, to constricted possibilities, but more likely to painful mishaps. I hear the words in my head, I catch the logic of the response to do the thing, whatever it is, that is already written into the way I have identified the problem at hand as a problem, not a consequence of contextual conditions.

I have often felt that to save my children, I have needed to brutally banish the scripts of modernity's logic from my bones. It is easy enough to learn new scripts, follow new recipes of the latest parenting trends. It does not take long to notice that they are well-meaning and potentially insightful.

Still, as programs or manuals, they were forged in the same factory that disconnects me from the realm of possibilities I am nourishing — or trying to. I did know that I knew in my belly, but there was no permission to attune to the ancient mothering within my blood.

Decades later, the learning in my belly has entered my skeletal marrow, and my spine is louder than the teleprompter. When my stepson was recently struggling in math, I was more inclined to gather a family meal, bring in the big siblings, and ask him to take care of my dying mother for the weekend than I was to punish him into getting better grades.

The gathering of the family is a show of care; the big siblings helping is a way to not be alone in the situation. Taking care of someone who is dying is a way to get a keen peek into the preciousness of each day, honesty, and love. These things are more likely to build the soil into which his morale can muster the courage to face his blockages around math.

Our household needed that medicine. The whole family needed it. The math crisis was just an opening, a very specific opening and invitation to do the "next best thing" or to meet the "relevant next."

But which "relevant next," and which "next best thing?" That is where the practice of spotting and stepping aside from insidious cultural habits that serve the ghosts of modernity comes in. It takes constant vigilance to try and stay out of reach of those ghosts. They are perpetually whispering, seducing, coaxing, justifying their own ongoing relevance.

There are rules. And sometimes, in order to commit to those rules, other rules have to be broken.

Will the ancient rule that binds parents to the well-being of their children and the communities into which they will bear children be overruled by the gluttonous appetites of individual success and wealth?

Is each child not a harbinger of possible futures?

Fig. 4. Vivien Leung. (2026). [watercolor]

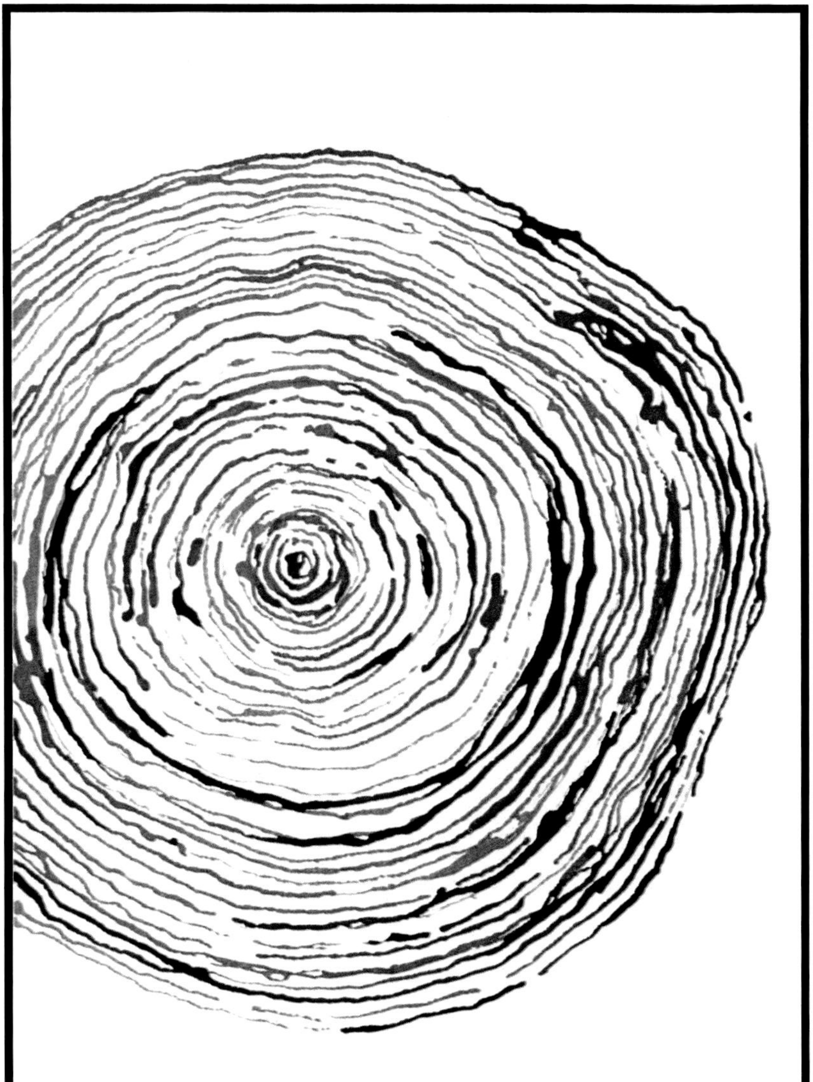

Fig. 5. Vivien Leung. (2025). [watercolor]

matrix

> ***Yes, of course it hurts*** *when buds are breaking.*
> *Why else would the springtime falter?*
> *Why would all our ardent longing*
> *bind itself in frozen, bitter pallor?*
> *After all, the bud was covered all the winter.*
> *What new thing is it that bursts and wears?*
> *Yes, of course it hurts when buds are breaking,*
> *hurts for that which grows*
> *and that which bars.*
> *(Boye, 1935/2014)*

Dear Mama,

You are gone now, but I need to talk to you. There are some things I never said, questions I forgot to ask, and I wish to sit in silence with you. I am trying to set you free, set your mother free, set your father free, set their parents free, and set my grandchildren free one day. We have to find our way out of these habits of hurting each other. We have to find our way out of the expectations of individuated success and failure. We have to learn together how to be in a world we do not yet have access to.

Mama,
we have to break the matrix.

I am of you, descended through your body, your habits, your bad relations with your father, your wild spirit, your ferocity.

And as your daughter in a time of paradox, I say again:

Mama,
we have to break the matrix.

The word means womb in Latin. It came to be used to refer to a pregnant animal. It means mother. The film by the Wachowski sisters brought it to mean the "source" of a simulated reality.

In our mother-daughter-mother-granddaughter bonds, a river current of continuity meets each new generation and pulls us into habits too deep to see. Habits formed along the way. Habits that kept us alive, perhaps, but at the expense of life. Habits of devitalizing. Insidious habits of isolation.

Every generation has contributed to the simulation ... it's a very rich story. I cannot repeat the program. The task of being alive now is to find a snag in the illusion and tear it open. Reveal the breath and water of being of life.

Matrix also means: table, board, index, fascia, chart. It refers to a laid-out array of source materials.

Oh, I see it now: the mother, the table, the fascia ... it's all coming together.

The tissues of our beings are pulled taut between the wombs of ancient times and the tables of today. I am not calling you wrong. I am not disavowing your beauty. I am not throwing away your gifts. I am not mocking your wisdom.

I am saying:
Mama,
we have to break the matrix.

Forward and backward in our intergenerational stories, this thing we bring to each day. Each new generation must show learning. I am saying I cannot be in the relay race to deliver my ancestors' double binds and scarred hearts to the children who need to have free hands to do the work of a new orientation.

I cannot tend the hiding places, or the protective shields, that are distanced from life and death and love. Now we tend the infection. You are gone, but not gone. I still feel you speaking through my hands, see your imprints in my home. I am your embassy now, and the rules of our conversation have changed.

As such, I have to hold this carefully, knowing you have built the castle of your familiar with your blood and tears, and those of your ancestors. But now, we are going to break the matrix.

The river turns when wolves rearrange its banks, and the deep waters pull into other gravity. The table and the womb, the source and the simulation. It must be painful to be born, and I know it is painful to give birth. It must be hard to crack out of an egg.

As I held you in your moment of death, I saw that too was not easy.

Mama, we do this together across time and dimension, on every table, in our wombs, in the index of being, all the way to the source. We break the matrix of yore to revere the matrix of what matters.

It begins between you and me, Mama, at the table.

mattering

When I say I trust you, what I'm saying is that I believe I matter to you.

And that ... even if we blunder, we will sort through the confusion and take care of each other.

But if I am mistaken — and I discover that I did not matter to you — that pain is among the most excruciating a human being, possibly any mammal or other species, can endure.

To learn that the relationship I thought I was in is not the relationship I thought it was is a profound misunderstanding of the context, of cues, of communication. It means I misread my surroundings in what historically amounts to a very dangerous way.

To misunderstand contextual relationships is, for all organisms, deadly. Survival is dependent on learning to be in particular configurations of interdependence. If a mouse cannot read the circumstances for predators, it will be food. If the mouse successfully reads the circumstance, it may find food.

Misunderstanding the context is fatal. It matters.

So it is easier to assume I don't matter to you. The Machiavellian logic of "what's in it for me?" births the assumption that you will abandon our relationship in the blink of an eye. I keep my tender underbelly safe from what appears to be inevitable danger. The cost is high. The price of this is isolation and a world of mutual suspicion and exploitation. Distrust is a healthy response.

What matters?

We speak of complexity, interdependence, and internationality, and the vocabulary is distant from the intimacy of a heartbeat to heartbeat relationship. Speaking of entanglement impersonally is a strange split.

How could this be?

The complexity is that there is no way to simply choose to matter to one another.

When it comes to the immeasurable mutuality of a living world, where what matters is the continuation of living organisms, the zoom in and zoom out of dedication is dizzying. The bacteria matter, but the mammals matter, and the reptiles matter, and the insects matter ... the meadow matters. And yet, if defining what matters requires parsing out of the larger among-ing of these organisms, conflicting matterings surface. What matters in a chess game is not at all like what matters in a meadow, an ocean, or even perhaps a human society.

Eachness is caught in infinite expression of the collective lifeforms moving, learning, among each other.

When I try to protect my children from the dangers of technology by putting defined limits on their usage, have I also condemned them to a world of secrecy? Each day is awash with moments to consider what matters. If our household is successful in holding up these protective limits, and the kids go to friends' houses where the access to technology is not measured, will we lose communication with them in other contexts? I have to include in my mattering about technology the extension of how our communication unfolds over time as the dangers I am tasked to protect my children from shift contexts.

What matters is that we are able to go through the difficulties of life together with communication and learning. Given that baseline, how might I shift my expression of why it matters to limit consumption of technology, because that matters too?

If we don't matter to each other, and if not mattering is treated as not mattering, then a discussion of complexity floats on the surface of experience, confirming distance through the professionalized concepts while shielding the taste buds and tear ducts from the ideas. They say things like "what is measured is managed," and the subtext here is that what is measured matters.

But does the logic and mathematics of measurement hold the necessary variables of movement that allow for life? Or does that not matter?

The embodiment of the conversation becomes an embodiment of a sterile, boundaried vocabulary, which ironically lands in a fermenting brew of epistemology — altering the moving and learning of the inter-life to include the punctuation of a numb grid.

It enters life, becomes life, carrying a repulsion of merging. The story of hold-back and walls becomes intrinsic in the sensuality.

When speaking of values — which is a muddled word carrying a fusion of economy and emotion — the underlying messages speak to what matters. When someone says something is valued, what they mean is "that thing" matters. What matters is what is valued.

My friend's brother escaped to Poland out of Ukraine and also managed to get his stepson out, but before he got out, he was shot. He went to Poland, received medical treatment, and then promptly returned to fight where he felt it mattered and he was needed. This is a common story of people who leave dangerous situations for safety, but in the process, they also leave what matters to them. They often return to help where they can.

How broad is the scope of what matters?

Because surely, if the mattering is limited, those limits will render Venn diagrams of mattering that exclude. The shadow of belonging is the edge, where the inside is identified by contrast with those that are not part of the belonging, by non-belonging. The cheapest form of belonging is exclusion.

Identifying "us" by describing (and vilifying) "them." Easy, fast bonding built on other-ing does not nourish the ecology of life-ing.

Will we cover for each other?

That is what infinity requires. No hold back. The world is inherently unsafe, and shallow promises or performative safety measures make me uneasy. Safety is perhaps best described as a relationship in which I can be both wrong and loved.

Do I matter?
Do you matter?
What matters to us?

Fig. 6. Rachel Hentsch. (2026). [watercolor]

digesting the grid

I invite the notion
that all experience is embodied and somatic.

The abstractions of modernity have been embodied ...

Watch the anxiety that arises physiologically when plans, metrics, and outcomes are removed from "systemic" projects ...

The issue is, in fact, that the grid has been lived into, alongside and marinating in the grace of life.

embodiment

It is not possible to have any experience that is not embodied.

We have eaten too many abstractions. I imagine they may have imprinted cubistic impressions on our intestines. The appetite for grids and more grids, methodologies and more methodologies, measurements and more measurements is that of a hungry, insatiable vampire.

The point is our bodies are holding the abstractions now.

Watch what happens when you want to hold a meeting with no agenda, or do a project with no declared outcome, or set out on a task with no goal ... you will see the anxiety emerge immediately.

Anxiety is embodied. The need for control after many generations of linear logic is embodied. The somatic is inclusive of the false separations, including the separation of the somatic from anything else.

Breathing, digesting, heart beating, cheeks flushing, fingers fidgeting — embodied gestures of the experience of a conversation. Eyes flickering, brain hopping, adrenaline-triggered — embodied responses of the experience of doomscrolling. Every experience is entering the body and the entire cognition system, within the complexity of culture and language.

Airplane food may have less wholesome nutrients than a farm-to-table meal, but the experience is just as embodied. Algebra class may have less attention to the placement of your arms than a dance class, but it is just as embodied. A trader on Wall Street may spend their days on phones speaking only of profits, but this experience is just as embodied as if they had been forest bathing.

The hardest information to shift in the verbal-conscious perception is that which is actually lurking far beyond accessible data, and in the fascia.

There is no way to select particular ideas and alter them on demand when they live in the spine, the belly, and the fingertips. There is no way to make a direct corrective to the way a body holds story, holds intimidation, holds fight, holds dreams.

To reclaim the body's knowing is to attune to its responses in every moment of the day. Imagine each hour is an embodiment workshop, no moment is separated from breath, from heartbeat, and from the sensorial information that enters an already swirling learning sea of other experiences.

Doing taxes is a somatic experience. What do you notice? Where is the clutch, the itch, the dryness?

Likewise, making a pot of soup is an intellectual endeavor. Full of ideas from many directions. What is the history of the stock, what is the culture of which knives, which chopping boards to use? What are the politics of the dish? What is the history? Different cultures have different soups.

It is all ritual, all embodiment, all intellectual, all cultural, all biological, all communication. Every moment is a bodily event, an intellectual event, an emotional event, and a spiritual event.

The punk rock is not to separate these.

The punk rock is to make soup for someone.

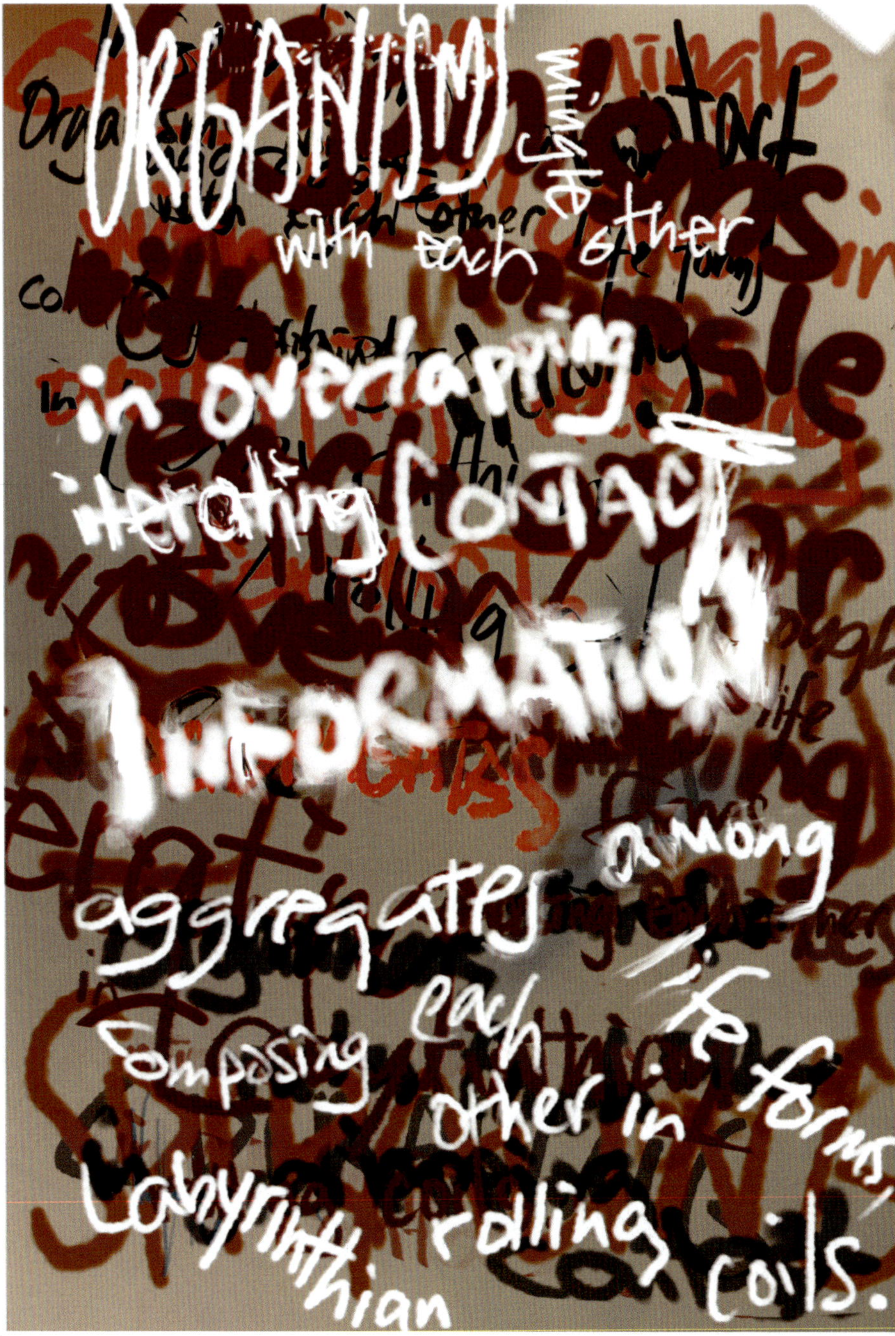

Fig. 7. Vivien Leung. (design), Nora Bateson. (concept). (2026). [digital]

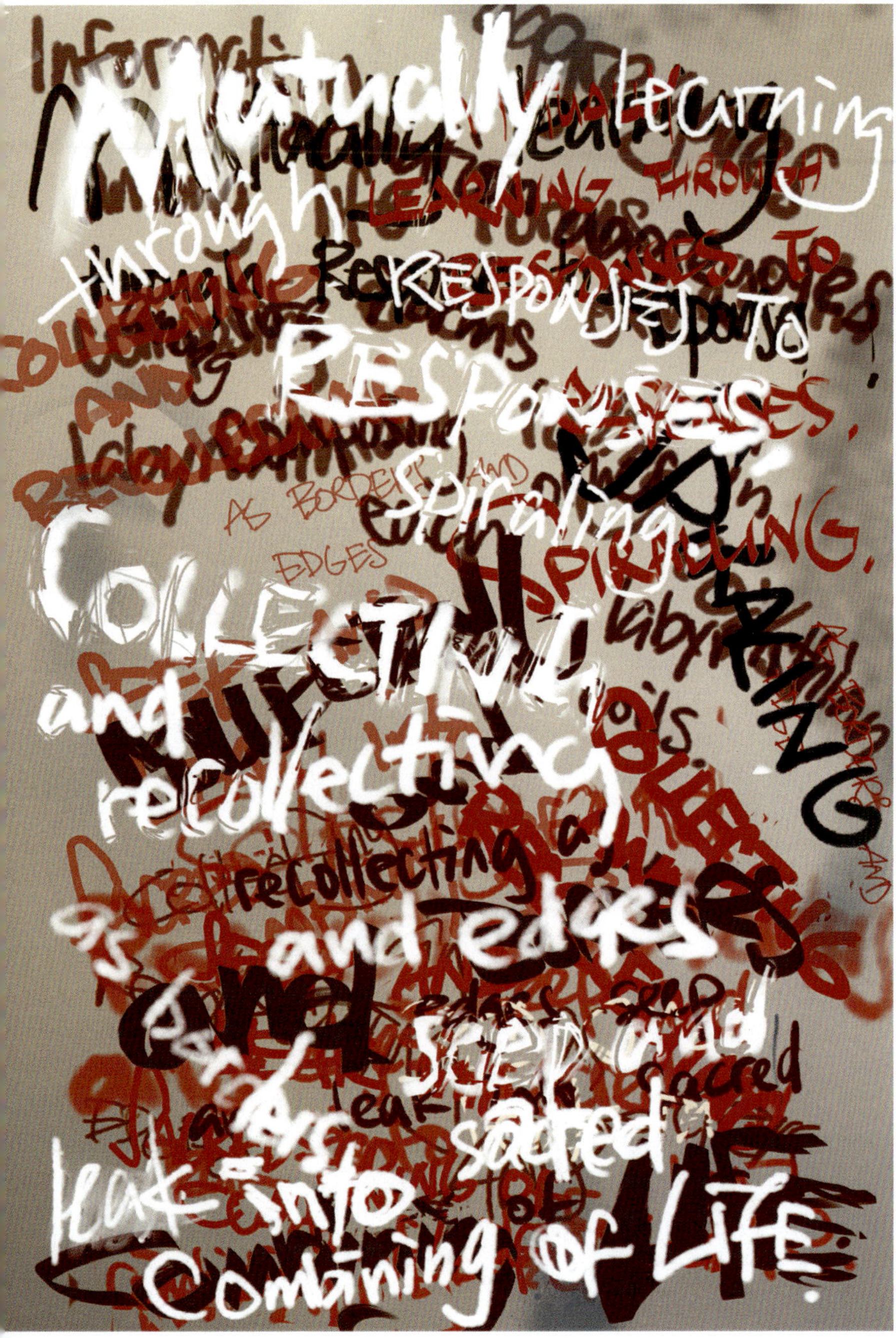
Mutually learning
through
RESPONSES
spiraling
COLLECTING
and
recollecting
and edges
as borders
sacred
leak into
containing of LIFE.

Fig. 8. Rachel Hentsch. (2026). [mixed media]

chewing ...

do not eat the compost

Any attempt to discuss the infinite, ever-recursive effects within life's complexity reveals a paradox: the language most capable of holding the sacred processes of aliveness may be found everywhere — except on the page full of words. The movement of ecological processes is not translatable into industrialized grammar and understanding. This is a double bind: using language where words so often confuse communication more than they clarify. Yet language becomes the necessary and inadequate means through which we attempt to express a shift in perception, a shift in how we speak about what is already there to experience in a flower, a song, a bowl of soup, a mother picking up her child.

Words lay claim to solid ground, even in our effort to liquefy them. It is difficult to communicate about perceptions for which there is no language, and equally difficult to form language around that which is imperceptible. The habits of perception and the habits of language are merged.

Existing language gives an indication of the type of perception that has been favored and entrenched within a culture. That we do not have a word for the infinitude of movement in life reveals a gap in perception of that infinitude. And yet, without perception and language to express it, there is a danger that this vital participation in life is obscured. What a "thing" to have no word for in scientific English.

Other cultures have named this life-giving phenomenon. *Tao*, from Chinese philosophy, and *Wakan Tanka*, from the Lakota tradition, are words other cultures have used to approach this vastness. Even without language, infinity murmurs through us.

Transcontextual anticipation: When I play "stick" with my dog, I sometimes tease him by pretending to prepare my body to throw the stick in one direction. My dog reads my muscles, he is watching the movement that precedes the throw. With excitement, my dog sometimes runs ahead in the direction I have pretended to throw the stick. We are both entertained by this game. His movements arise in anticipation of mine. I am playing with his anticipations. We are both inside the shaping of what has not yet happened.

All organisms are doing this all the time, through multiple contexts of communication simultaneously. The spots on the deer anticipate the dappled light of the forest. The squirrels gather nuts and seeds anticipating winter, and the next generation of forest anticipates the squirrels will forget where they hid their acorns. When we train our muscles and strengthen our bodies, the possible uses of those muscles are myriad.

We may carry water, or carry a friend, we may build a house, or we may move firewood. You can use your legs to stand up, or to climb a tree, or run from a fire. The muscles are there in anticipation of actions in many contexts.

The same multi-directional anticipation is present in the physiological systems of systems of the human body. The eyes see regardless of what is seen. They are preparing to translate light and contour into contrasting sizes, shapes, and colors so that the seer can find their way. The digestion will do what it does regardless of whether you are driving a car or taking a walk. Organisms are composed in countless ways to respond to the variable conditions that may surround them.

With transcontextual anticipation in mind, it becomes possible to marvel at the fact that every living organism exists within ever-shifting contextual potentiality to respond to its surroundings, each within its own limits (which are always evolving). This configuration of ready-ing allows life to continue in mutual learning as it changes. What appears as prediction is an effect shaped through each organism's histories.

When you open the fridge you may begin to salivate in anticipation of something nice to eat. The body hypothesizes what might be eaten. In the hypothesis reveals what you have eaten in the past; culture, habit, season, and addiction are all present in the imaginal moment of opening the fridge. The prediction of what may be experienced is carried through the senses, skeletons, nervous systems, and digestions of each.

What is it possible to perceive, and what will the limits of that perception lead to? Cause and effect more than loop here. By contrast, the rigidity and tight coupling of structures that have too few possible responses cannot meet life. If an organism becomes habituated to eating only one food, or to only communicate in one way, or to only use a few muscles, the ability to respond to life becomes too limited.

All organisms, including human beings, are responding to events around them with experience that reaches beyond specific enactments. Birdsong in the morning is more than the birds trilling their joyous song into the day, and yet that may be how I hear it. The birds are using their bodies, evolved over epochs, to make these sounds, which reverberate around the forest, signaling daybreak, safety, springtime, mating, location, and much more. The nervous systems of many organisms are soothed by the birdsong (except perhaps the insects and worms). Even human nervous systems register that when birds are singing, there is no immediate danger nearby.

But the causal processes of the birdsong are easily reduced into a particular purpose of "why the birds are singing" when, in contextual terms, the effects of the birdsong are widespread throughout the organisms within range of their resonance. Are the birds singing because they can? Because they want to? Or because it's spring? Or because it's morning? This initial causality is impossible to pin down, as is the edge of the "effect" of the birdsong. The birds are singing because birds sing, the other organisms are responding because their ears or other receptors of sound are able to receive the song. From this sensory stimulation, the organisms produce actions. The causality of the organisms' actions is in the effect: the birdsong illustrates the history of each organism's physiological capability to receive sound — communicating not just springtime, but the entire year's seasons by contrast to spring, and a message of the absence of predators and storms into the entire forest as it is home to both the bear and the bunny.

Rachel Carson's (1962) book, *Silent Spring*, evokes a particular terror in what might happen if the spring did not bring birdsong. The anticipation of birdsong throughout the forest creatures is fundamental to the forest itself, from the insects to the bears, and everything in between. This is just one example of one creature's expression of life in an ecology. Imagine the infinitude of expression in a forest, and that it is all ever-changing according to varying variables.

To be in a world so full of movement from so many directions and contexts is to be in constant action. Even non-action is action. There is no way to avoid each moment being an effect of all that has come before, and a causation of so much that is in play as well as a participation in what follows. The idea that a single action can be carved out and strategized into a singular effect is revealed as an error of too few considerations. The patterns of response to the world need to be like the patterns of movement in the world.

In human terms, a practice of perception and expression of these critical processes of life is a condition of continuing to be alive.

Fig. 9. Rachel Hentsch. (2026). [watercolor]

Fig. 10. Noah Qwarfordt. (2026). [acrylic]

how clean is your kitchen?

Nothing is less fun than righteousness in the kitchen.

There are always people in my house — kids coming and going, some with partners — in general, it's a bit chaotic. We all have a hard time tracking the laundry and the fridge contents. The countertops are in a forever state of movement, items ranging from mail to groceries to spices and oils. Chargers go missing, bread gets hard, we run out of basic ingredients. Remarkably, the whole mess of us seems to find our way to the dinner table each day, where the places are set with care and the fractures of the day are soothed into reunion.

My kitchen is only ever clean for a nano-moment before someone comes in and starts using things, moving things, eating things, cooking things. It's a hub of activity that is always in go-mode. We do try to keep it clean, but the effort is not very satisfying since the entropy is immanent, permanently. I find when I can relax the part of me that judges myself for not being the perfect householder, that the crumbs and random cups in the sink also provide a sort of comfort. My kitchen is not a tightly run ship; there is nothing tight about it. While the cluttered counters might jangle the nerves of some folks, the meta-message is also conveying a place teeming with life and love.

Each passer-through who touches the toaster and the tea station somehow contributes to an endless storm of loving fingerprints invisibly casting a glow of affection. It's an altar, scattered and strewn in the remnants of our household's tangle of stories as each of us comes and goes during the day.

Sacred chaos of love and life is brewed here.

Of course, I have felt terrible shame at the disorder. As a mom and woman in the world of housekeeping judgment, this is a belly and bone response. I can't help it; my mother and my mother's mother's mother are all measuring me up, and I cannot shush their insufferable snarking. I long for the magazine-clean-featured-display-clear counters and everything perfectly-perfect kitchen. It's just not possible. I would have to either never sit down or become a tyrant to achieve the look of a kitchen unused. Not worth it. Our kitchen is in use, and it shows.

I remember my sister telling me that she invited her students into her home once a year and intentionally didn't clean beforehand. She said, "It is not a kindness to them to present the illusion that a house of an active, working, creative life is a clean home." The students, many of whom were women, needed to see the real thing, the little bits of life in play.

My husband is less tense about the need for order than I am, but only because, let's face it, when people arrive and they notice the beautiful chaos of our home, they don't immediately think, "that guy isn't keeping this place tidy." No, somehow the disorder always reflects upon the woman in the house. Maybe it's different in same sex households, but somehow, I sense I am always the one at fault for failing to keep the house immaculate.

I want the kitchen to be homey, relaxed, alive. I like the clues here and there that reveal someone's been having tea. The smells of recently cooked meals, and the ongoing slurping of the dishwasher let me know that people are being people-ish and gathering in the kitchen.

The best dance floor in our house is in the middle of the kitchen, but even if it were not, that is where the dancing happens daily. A wiggle while stirring, a little jig over to get the next ingredient out of the fridge, a twirl and a kiss by the sink while doing dishes — this is the dinner hour. I love the love in it. So easy, so free of charge, so much more fun than stressing.

I get upset about getting upset about the kitchen not being clean. With so many people coming and going, absolutely everyone has a way to slide out of responsibility for the mess. It's hopeless to try and pin it on anyone in particular. We are all responsible.

But conveying the flip side of that responsibility is not so easy. Cleaning up other people's messes is irritating. Everyone, of course, feels they do so more than anyone else. To declare a reason why this mess or that mess belongs to someone else is easy, but volunteering to be the one who picks up the slack is not. People start doing dangerous math, keeping score, making chore wheels, getting moralistic, preaching. Again, nothing is less fun than righteousness in the kitchen.

The buzzkill is second only to the ripping threads of relationships that nagging about the dishes generates in a family. If only it weren't necessary. But sometimes there is nagging. Sometimes there is scorekeeping. It's not pretty or graceful, but when the heart fills with it, sometimes the grudge needs to get out. Nagging never helps; it only makes everything worse. Nagging distances the kids, pushes away the spouse, and generally ends in lonely bitterness smeared across the family.

I remember one time asking my son to take out the trash, and he very, very, very, very wisely replied, "I will do it in a little while, as soon as I forget it was your idea." Because doing something nice for the household is a pleasure in and of itself, but it is one that is removed when the task is demanded.

This is not just about doing the chores; this is about how the future world beyond controls, grids, and mandates is possible. Clearly, we all need a lot of practice. Especially between the generations, the nagging and demanding have multi-generational habits of guilt-tripping in the language and tone of the invitation. It has taken me a long time to learn to simply ask for help, to give context without subtext. To find my way past the resentment of so many ancestors who served out of obligation.

I want to serve, but out of inspiration, together. I do not want my kids to be saddled with a choice between dishes and a guilt trip. I just want them to kick in, to see me clearing up and instinctively help out. "Keep dreaming," you may say ... but it's happening, and more importantly, the pattern of communication is shifting.

Perhaps my grandchildren will never know of these terrible double binds we locked each other into. Perhaps they will never know of parents who nagged, and kids who were guilt-tripped, and mothers who felt the shame of the cultural expectation that the clean house was on them.

How clean is the kitchen? This question becomes a barometer of a sort of ready-ing of household communication that reaches far beyond the kitchen and into the entire question of how we live together.

Do we maintain the joy and love in our relationship in enough ways to be able to see when the dishes need doing? This is not only about dishes, but about all of the work.

This is a new economy, this is a new culture, this is a new parenting, this is a new social code, this is a revolution against me-mine, this is a new politics. This is a respect not just for dishes but for how living organisms move from a zero-sum game to an infinite game. Unlike a game of cards, the game of dishes and food prep never ends, and it is the perfect place to practice shifting the communication into a nourishing that is ongoing, where the objective of the game is not to win, but to keep the game going. Today, tomorrow, and three hundred years from now, the practice of generosity of all parties in the kitchen is radical. It is seminal. Can we eat together without intergenerational, inter-gender, inter-cultural harm?

There is a family literacy of signals around who is in what sort of state at any given moment. Who is tired, who is sick, who is in a work call, a school project ... who is available to do the dishes? This relational literacy is not forged by force. It is fed by each day's communications about a thousand other things. It means noticing each other's lives. It takes a little time to find expression, to share the day, to voice the ins and outs of each of our situations.

Mostly, to dwell together, such that the relational circumstances are relayed, is a practice of mattering to each other. This practice is not about obedience, or control, or the bare minimum. It is about a generosity of tending that must be improvisational, not planned, not pre-determined.

This is the practice of living in loving complexity. This is a kitchen that is not exactly perfect, but is generally clean enough. The effect is not a linear cause; rather, it's but a swarm of responses to a swarm of events, some of which were in past generations. So expect uneven results.

Untaming the kitchen is an art form of shared delight. The practice in the kitchen is non-stop; every day, multiple times a day, food and kitchen time are there for us to remember how to be together. How shall we manage to keep the kitchen respectable without a rule of law or a schedule or a scorecard ... all of which obscure the contexts in the room?

The contexts are more than in the room. The contexts are in the culture that needs me to feel mommy-guilt and mommy-rage. I would rather not be communicating with my kids through either of those lenses. I don't want to continue the previous generational scripts of shame. I don't want to lace each request for assistance with an underlying assumption of individuation.

The future is households of many generations. Most of us are wholly unprepared for this. Some cultures still manage to hold back the isolation of individualism with a strong weave of family relationships. Others have succumbed completely to the lonely world of the modern self.

What if we break the matrix and dance in the kitchen instead?

How clean is your kitchen?

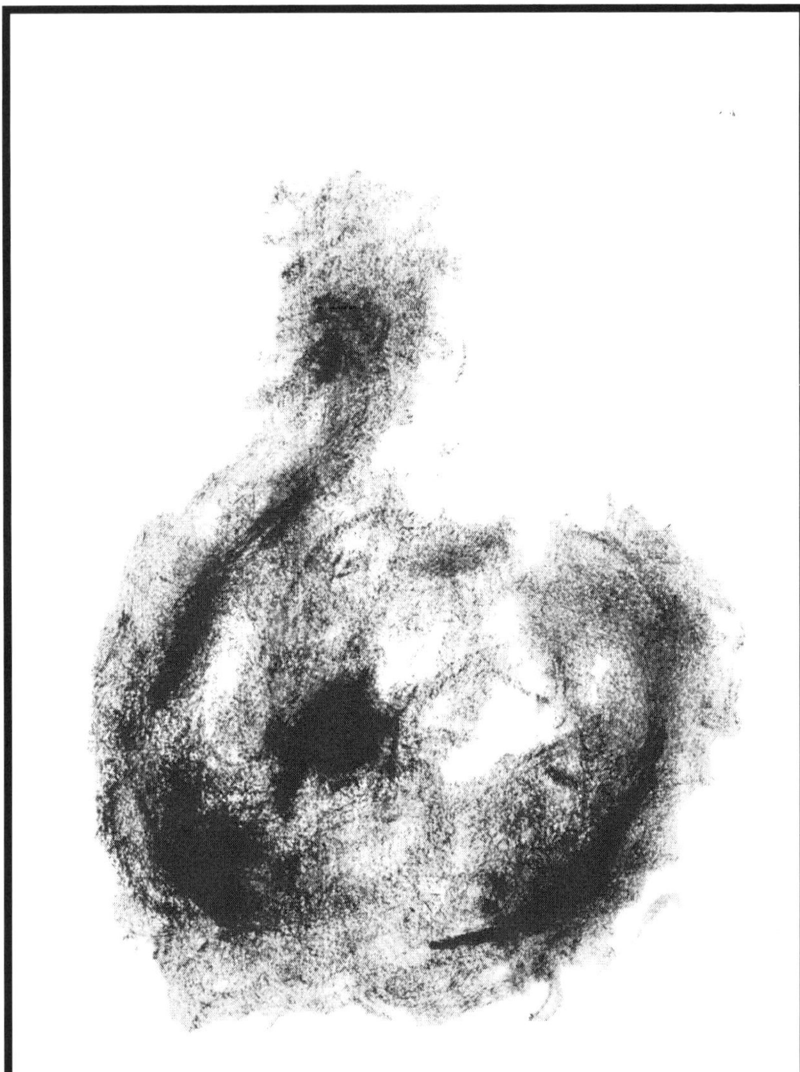

Fig. 11. Noah Qwarfordt. (2026). [acrylic]

ceremony

Breath and bells. The kettle boils. Mountains loom.

Flames pull upward and downward. Ground and sky. Herbs gathered, dried by sun, ground by stones. Seasons circling. Simmering water pouring over the herbs,

Rusty oranges of fire, pea flower blues of water, cinnamon amber of earth, whistles and burps of air frothing. Leafy fragrance in the steam perfumes the pores, the lungs, the nose, Clouds inhale.

Smoke, incense of burning things, sings into the steam. Their particles mingle. As ever, forever, ever more.

The herbs soften and open in the warm water, preparing to heal. Already the body anticipates this ancient medicine,

With two hands the cup is brought close. Lips filling with heat, throat welcomes, belly receives. And the body is fire, the body is water, the body is wind, the body is soil —

Minding the children of ancestors and future generations in a continuum of awe. Re-minding dimensions of their eternal dance.

Breath and bells and generations.

Fig. 12. Rachel Hentsch. (2026). [mixed media]

chewing ...

orders

Causality in interdependent systems is commonly described in terms of "orders." At present, to bring the world of systemic theory, complexity, and cybernetics into this discussion of communicational, ecological, and cultural change without the use of the term order would generate confusion. Yet, for those coming to this conversation from other fields, the term order also generates confusion. Personally, I would like to move beyond this language of orders in an effort to keep the language itself in motion. But first, it is important to address this language and its usage.

The term order comes from mathematics and has been adopted in different ways in a variety of cybernetics and systems theory contexts. As I use it here, it is an attempt at getting at the gist of thinking about cause and effect, beyond the first outcome. In this language, there is reference to first-order, second-order, third-order ... and so on.

Each order represents another zone of causality. This language is used to describe how higher-order systems can regulate and influence lower-order systems through recursive relationships, providing a framework for understanding the complexity and interdependence within such systems.

In the way I am using the term, first-order refers to a direct action, with the intent to achieve a direct goal. I turn off the light, the room becomes dark. I want to eat an apple, I eat an apple. The notion of first-order is useful to contemplate, as it is a reminder that it is strangely possible to imagine that an action can be removed from all other considerations and ramifications around it. The notion of first-order is what feeds the hungry ghost of control and mechanistic, measurable outcomes. Reductionism is seductionism ... and argues for the "simple, provable, understandable" truth of "this causes that." An action with a first-order goal is linear, direct, and correlative.

What is sometimes called second-order refers to the considerations: Why do I want to eat an apple? Do I really want an apple? In these questions, there is another layer of information referring to the causality of the wanting, and the apple, and the observer, or agent. One issue with this description is that, upon deeper consideration, nothing just happens at first-order.

In a world of interdependency, everything that is done and not done ripples into responses, and responses to the responses, from micro to global complexity.

The only thing that can remain first-order is the illusion of first-order.

This illusion is dangerously satisfying and easy to grasp after 200 years of regular doses of industrial logic. The notation of a seemingly first-order effect is only possible if the surrounding relationships are edited out of the spreadsheet. In a machine, each part has a role in the function of the mechanism, but in a forest, or in your gut microbiome, no organism has just one role; instead, redundancy is the ecological approach to inter-relational vitality over time.

When an event is located in the idea of a role, the effect is parsed into the confines of that narrow "role" relationship, and it can be seen as a first-order effect. But this is a tiny clipping of the larger shiftings. The organism, the event, and the parsing of it all require more than first-order description to adequately take heed of "consequences."

Complexity in life is often described through something like the butterfly effect, except, instead of following the effects of one event, (the butterfly flapping its wings becomes a storm on the other side of the world), it is necessary to attune to the event shifting the course of entire ecologies; playing out simultaneously through millions of organisms, cultures, and ideas — change changing change, changing changes ... each transform reshaping the next, spiraling onward into recursive dizziness. When an event meets entities in relationships that exist through redundancy, the correlation goes everywhere, without end; again, no organism in an ecology has just one role.

Fig. 13. Vivien Leung. (2026). [watercolor]

the belly holds the world

All the places and people, the awe, and the pain churn in the mid-torso orb of life. The gentle, churning gourd bulges with juicy stuff as it is home to many generations of ghosts who inhabit each living being. What they found in the world they lived in, how it met them, all stirring and digesting.

Infinite information in-formed in tubular labyrinths writhes. The belly is an ocean of times folding together. Somewhere in the roundness is the first fire, the first crops, the feasts of yore, and the famines. Envy and fear live here in the same liquids as inspiration, infatuation, and the belly laugh. Softness. A pillow for a beloved.

A lot of organisms, scientists can't agree on how many, live in the belly. The organisms also harbor generations of ghosts inhabiting them. The organisms communicate with plants and dogs, viruses, and love hormones. They draw us together for coffee, cake, tea, and sweets. Bacteria love caffeine and sugar. In our belly, we have a nonstop party of history and biology that has not forgotten that which the conscious verbal has lost. The belly digests words into hypotheses. Metabolizes nutrients into metaphors. The belly calls in ideas and paints them into our blood.

The belly has learned when to be afraid; it has found that it's okay to eat artichokes; it taught us fermentation. It communes within the field of the body, the context, and the overlapping timelines of life of which we are. The belly is the squishy core of primordial possibility. Bubbling and burping, a bellyache is signaling both distress and indigestion. Blurring emotions with potions, our bellies speak in groans and gurgles of millennia.

Bellies are alive.

sensitizing

The research that needs to be done to bring science to what most ancient cultures have always known, has barely begun. It is all well and good to say that everything is interconnected, but those words are too easily expressed, while the depth of the ramification of that interconnectedness remains illusory.

The research around microbiomes is beginning, the research around mycelia is beginning, the research around the way breast milk responds to the baby's needs is new, the research around how trees communicate and nurture one another is beginning.

Significant to each of these new fields of scientific study is a new common sense, a new logic to live into.

Where is the edge of the mother?

Where is the edge of you?

Where is the edge of the tree?

For too long the research has doubled down on the cultural habits of separation, identification, and isolation of subjects. With this separation has come the obscuring of, and the erasure of the vital edgelessness of what being interconnected really entails.

Your senses are not even yours.

What does it mean when we say we are "making sense" when sense making is taking place through multiple organisms?

The following quote from the marvelous biologist Lynn Margulis brings me to my knees in astoundment.

> There's one thing that's very clear. Our sense organs: this means our taste organs, our balance organs that keep us upright, our rods and cones, the famous visual sense organs of the retina of the eye, the cilia of the ear, that means our hearing, well these sense organs, in general, are modified cilia. So that's the first point, and the whole spirochete idea is that cilia themselves come from once free-living bacteria. …
>
> The way I think of it is that sensory perception originally evolved in these bacteria. Now that's quite true. We have spirochetes and other bacteria that are sensitive to light, that are sensitive to mechanoreception, to touch, that are sensitive to all kinds of chemical stimuli, including glucose. And we know that glucose has big effects on the brain. And I think if we think of ourselves as being inhabited by organisms, or once free-living organisms, now completely tethered organisms, that have their own needs for food, for stimulation, for all kinds of sensory proclivities that I've just mentioned, that we can find a continuity between living consciousness, awareness, sensitivity at a microscopic level, and at a macroscopic level. I think that throughout our biological history fusion of different organismal types has been much underestimated as an evolutionary phenomenon. And we see this in social organization in people. (Margulis, n.d.)

Fig. 14. Rachel Hentsch. (2026). [mixed media]

chewing ...

transform

The word "transform" is curious, and central to both my father's work and that of Charles Sanders Peirce (1931–1958), who recognized the importance of the way something is changed by perception. In my father's terms, a transform is defined as "(coded versions) of the difference which preceded them" (Bateson, 1979a, p. 109). I usually translate his definition into more grounded language: perceiving stimuli through the coded experience of having perceived other now-familiar stimuli. The familiar is the filter through which new perception is ingested.

Later, biosemiotician Jesper Hoffmeyer (2008) spoke of "semiotic scaffoldings" while addressing the same issue. What is perceived is not what is there, but what the organism is habituated into perceiving. Peirce called these "hypoicons" (1931–1958, CP 2.276–2.277) or sorts of icons or signs that refer to objects through similarity or likeness. Icons are imbued with meaning by previous contextual storying; they are otherwise irrelevant.

Most people who visit the Vatican no longer know the stories of the iconic figures painted on the walls and ceilings. But in the 1500s, when the artists were commissioned to illustrate those stories they were recognizable and relevant to most people. Today, people are more likely to recognize Marvel superhero characters and their stories than those in the Vatican.

In each of these explorations of the issue of transforms or interpretations, there exists a wild card: these ripples overlap, amplify, and transform each other, continually generating new ripples of ripples. It boggles the mind.

If calculating these changes mathematically, it might be said that we are looking at nth-order change, where n = infinity.

Effects revealed through time-folding phenomena are sometimes referred to as nth-order. "Nth" gestures toward an unspecified, extraordinarily large ordinal number, but even that term falls short of communicating the infinite, incalculable, unending, transcontextual nature of these effects. The term is imperfect, but it will have to do for now.

Fig. 15. Rachel Hentsch. (2026). [mixed media]

Fig. 16. Vivien Leung. (2026). [watercolor]

before times

In a memory belonging to some other generation, I can still hear the seasons,
The dust of summer and the cool sweat of autumn smell of time —

In this reverie there are no beeping things.
There is breath and muscle strain, a quiet that is everything but silent.

Chattering leaves in the wind, roosters, water cans, bickering old couples, plaintive cries of a baby, the neighbor's dog nurses puppies.
An old engine sputters. There are fires at night. Old men talk, smoke, and drink.

The thought of going somewhere is not minor.
It is a whole day, a whole week, a whole thing.

A bell around a goat's neck clanks.
The sun is hot, the rain runs in vein-like streams down the road.

Minutes are longer in this realm.
The quickening is earthy.
Every organism is constantly shifting toward the next season.

continued ...

I am trying not to forget what it is like to walk on the dirt road,
The crunch and slide of the small rocks — gravel shoveled by relatives.

Clocks ticked. I have coins in my pockets.

Life was not ideal.
People were still people, all mixed up and broken in a thousand ways.

To meet this mess is to perceive its still point, not spin out.

It is coming fast now.
These are the before times.

A buzzing in the spine, at the back of my skull,
a rapid strobe to the nervous system —
a sickness of timing.

The day arrives digitized —
jittering in my eyes, formatted and sized.
Everything has been optimized. Strategized.

Still … stillness is waiting somewhere.

Slowly, knowing the rhythm section is tidal, lunar.
Heart is thumping, both both both both,
dancing in multiple time signatures.

Nostalgia is not good, not bad,
just important not to forget how to do things …

Like how to be.
Watch a tree grow.

How to read a handwritten letter, and how to write one.
Making shoes is knowing feet and knowing leather,
knowing tools, and knowing the ground to walk upon
There is more to it than ordering a size 8.5 online.

I am reminding myself over and over to remember this
dream in every detail.

The belly button is there telling us there is an old, old
connection, plugging us into another source.

Let the ancient pulse soothe and restore the jangle of this
era, settle the soul, and inform movements from the bones
and roots of eternity.

Fig. 17. Vivien Leung. (2025). [watercolor]

forming

Don't freak out.
We are only freaking out.
Finding a way
Bending the way —
Like —
Those boxes with the swords poking in.
Shaping the contortionists into akimbo zigzags,
Limbs bent around the sharp ends ...
And inversely: fingernails in wood grain —
Holding shredded supports —
Like —
The creeping vine that finds its way up a flat concrete
wall, tiny gripping fibers seeking every micro foothold.
Reshaping in the context.
Remaking in concave, converse constructs.
It's all impossible improvisation.
Don't raise your eyebrows.
We are on stilts. Still being rebuilt.
Memory foam that forgets.
As it warps into the next, forgets,
As it warps into the next.

Fig. 18. Rachel Hentsch. (2026). [mixed media]

chewing ...

not all onion skins grow up to be conversations

I invite you, the reader, to travel along a serpentine path — an exploration of communicating about communication itself. Opening the study of communication in biological processes highlights the difficulty in attending to ecological communication. Both communication within the living world and our communication about it are vulnerable to the mechanistic habits embedded in languages shaped by industry.

The integration of semiotics or messaging into a fuller view of the living world necessarily reshapes our understanding of living context, even down to what is too often reduced to "simple" energetics or mechanistic function. This book explores the living world as inseparable in its aspects. I cannot separate biology from bioenergetics, and I will not try to. Instead, I will treat them as life does: in integration.

The times are desperate. So much needs to be done. Is it possible to act, to give action, to make action differently, such that both the immediate and the infinite are touched with vitality? If so, how shall we speak of it?

The ghosts of the industrial, colonial logic of modernity crave mechanistic strategies that are destructively out of sync with the interdependency that sustains ongoing aliveness. These ghosts also obscure vast realms of possibility with tempting illusions of practicality. Hungry for control and eager only for traceable causal lines, these ghosts seek their likeness everywhere.

Living under the skin, under logic, under culture, and under the hood, the habit of mechanized interpretation has haunted and atrophied the muscle of perceiving how life actually operates. Problem-solving has been reduced to seeking a single or mono-cause in a multicausal process, treating reductionist causation as the best approach to understanding.

Once found, the "viable" solutions to these "mono-causal" problems are locked into parallel, reductionist instructions. This is a central obstacle toward systemic change.

By contrast, when one eats an apple, the apple does not enter the belly and generate the ability to go around "apple-ing." Rather, it is digested and metabolized into energy that might become fuel for the person to wash dishes, or walk in the woods, or dig a vegetable patch. The energy from the apple is not easily extracted from the body's other sources of energy, nor can it be predicted or correlated to what actions it will nourish.

The apple goes in, and it transforms into what could be an infinite range of expressions of its gift of sustenance: perhaps the sustenance fuels a conversation that sparks a friend's new idea; perhaps it supports a walk in the woods that calms the nervous system and allows for sleep; perhaps it provides the strength to dig a vegetable patch, which never produces vegetables but instead becomes a learning about the difficulties of gardening; perhaps it nourishes dreaming during a nap; perhaps it is burned in the concentration required in writing these very pages.

We can't separate the source from the expression without destroying the continuity that makes it alive.

In each of these scenarios, the apple has offered something to the person, and the person has offered something to the world around them — but how that offering will find its next form of action is almost completely impossible to predict. And if the apple were eaten by a horse, what then would it become?

Let us consider such movement of transforms while imagining actions to meet the needs of this time, and the premises of cognition within those actions. Within the logic that has fashioned vast entanglements of overlapping, co-generating crises — embedded in both language and the culturally sanctioned proposals for action — the Möbius strip of self-fulfilling causalities will only continue to coil.

What is needed is an approach informed by the way life learns and changes. Such an approach will not resemble the familiar calls for action shaped by decontextualized information aimed at supposed, pre-determinable outcomes. The authority commanded and demanded by boards of directors and other committees, locked into direct corrective actions, makes it nearly impossible for them even to propose responses whose outcomes cannot be predicted. While the science of complexity and ecology is substantially more equipped to offer other avenues of action, even these sciences are subject to the undercurrents of culture that demand measured, context-specific solutions to narrowly-defined problems.

What is needed is action toward vitality — vitality-creating relationships in families, in food production, in education, in all aspects of life, for both humans and other organisms, and their ecologies. Vitality does not emerge through linear means. Systemic vitality is not graspable, modelable, or predictable; aliveness itself contradicts engineering and resists being reduced to a program or plan. The danger is that vitality becomes a banner under which we smuggle in the same metrics, optimizations, and instrumental logics that have already narrowed our sense of life. Health, function, efficiency, resilience, all can become endpoints to be engineered, tracked, certified, as though aliveness were an achievement.

Rather, the unknowable aliveness of information and response is perhaps the most radical basis for action — not because it is grandiose, but because it is so subtle, so indirect that it may appear to bypass the very crises at hand. A solutions-driven culture is drawn to perceivable relevance, toward justifying action, securing funding, and maintaining coherence within familiar frameworks. The craving for direct alignment between problem and solution is culturally condoned and embedded in existing institutions. The habit of achieving clarity of strategy is steeped into the conditioning of culture, and into the very ways we communicate and perceive the problems themselves.

Living systems move in other ways that will appear unreasonable to the logic of industrial practicality and incoherent to the conditioning and logic of control. There is increasing urgency to practice perceiving biological and systemic movement through the communication of responsive action. With practice, humans may become less destructive toward one another and the ecologies we inhabit.

The violence of historic associations lurking in the Old Testament and other texts, proclaims the duty of "mankind's progress" and the mandate to transcend nature. Recently, this acceleration of "progress" has been a central investment in the totalitarian technological trajectories of AI, robot technology, and the cyborg dreams of transhumanism. The need to meet the analog aliveness of the biological realms, of our bodies and all the organisms on this beautiful blue orb becomes ever more crucial.

The path of the apple is not unlike the path of compost — its transformation moving through countless shifts in form. The transforms are unpredictable, multi-layered, and relational. Compost nourishes the soil in which vegetables grow to become food but compost itself is not dinner; many transformations must occur before that food reaches the table.

Once grown, the food meets intergenerational recipes and is served to families and gatherings in varied cultural configurations of culinary etiquette. At the meal, prayers are offered, conversations unfold, relationships are honored and dishonored. From a decomposing onion skin, it is impossible to foresee the conversation at the table, and where that conversation may carry the relationships.

When I was working on my book *Combining*, I found myself writing these words:

> While compost is necessary for the growing season, do not eat the compost. Wait. The eggshells and onion skins bear little resemblance to the conversation at the table where the meal they fed feeds you (Bateson, 2023, p. 350).

Don't eat the compost.

It seems a rather obvious statement, and yet most efforts to enact change in our troubled world stumble into this same misstep in the order of operations. The takeaway, the outcome, the deliverable, the wish, the vision, all rely on a linear immediacy of likeness between identified problem and solution. The broken education system appears to need new curricula. Social injustices of poverty appear to need financial solutions. Particulate matter in the atmosphere appears to be a result of carbon and, as such, requires solutions having to do with carbon usage.

The compost teaches us something else, a difficult reminder in a desperate time, calling for another way of thinking about how change happens. Compost does not grow into a meal; it becomes soil. Don't eat the soil. Soil does not become a meal; it nourishes seeds. Don't eat the seedlings.

The transforms are many between onion skin and meal. And not all rotting onions grow up to be conversations; some rot in the ground, some are lost in dumpsters, some become dust and blow away, while others are in the kitchen buckets of home farmers who use the food waste of yesterday to nourish the soil of next year.

The road from onion skin to soil fertilizer may seem like a simple causal pathway one could follow without getting lost. But the next leap, from growing the food to gathering the people who are dining together — their relationships, their communing, their culture into which the reincarnated onion skin is precipitated is another matter. The onion is present in altered form, a transform of its onion-ness, in the gathering itself and in the memories those people may share of that evening meal. And what was in the soil that nourished the onion?

The catch is in noticing the expansion of the kinds of change that are taking place simultaneously. What happened before the onion was an onion, and after? Tendrils of story run in many directions, producing conditions, and conditions for conditions, that allow all sorts of life to unfold, not just a lineage of onions or even of gardening.

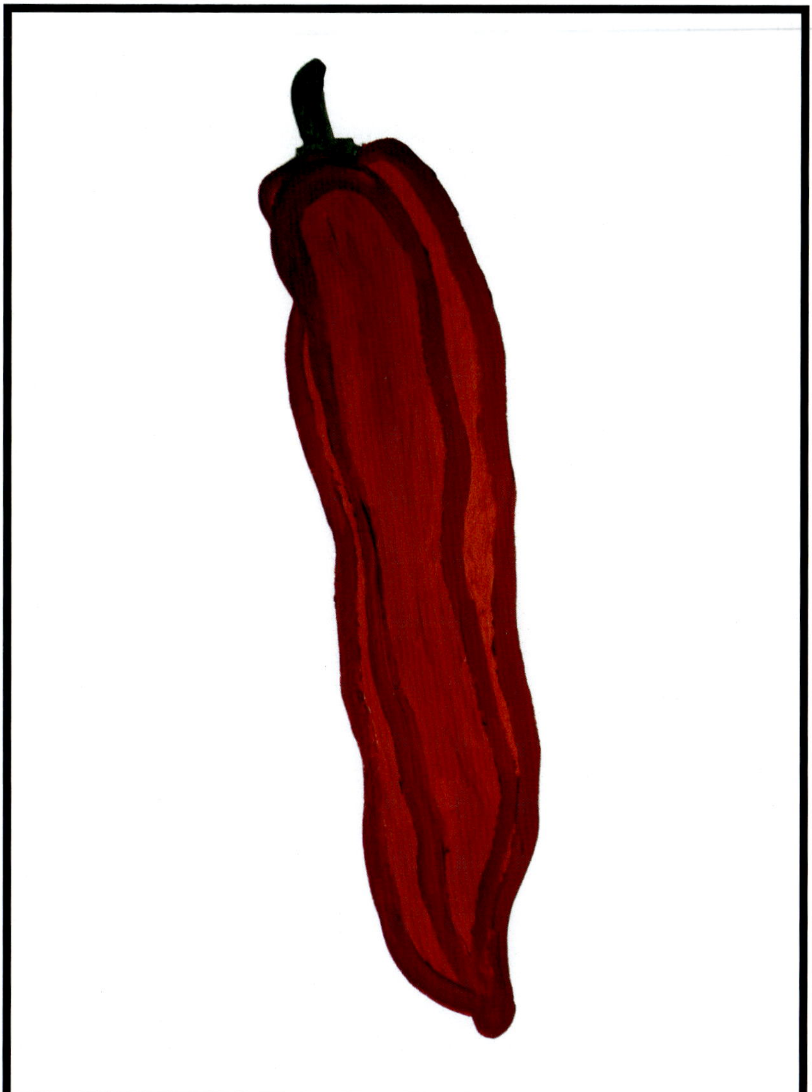

Fig. 19. Noah Qwarfordt. (2026). [acrylic]

waxing and waning

I remember when my belly was a child's tummy. It was a rounded ball that hid my small feet. Toddler belly.

I slept on my father's belly-rotunda while he gave lectures. He coughed a lot, and it bounced. Father belly.

When everything was tears in child-drama, the squishy warmth of a grown-up's belly messaged that the world is okay.

In the leggy years, my belly was just a button, no extras. No waist, no hips, just a skinny mountain girl. Knees and red hair. Child belly.

Then hips arrived, and crop-tops. It became a waistline, sassy in low-riding bell-bottom jeans; there was sloppy sex. Regrettably, I had no idea how lovely that belly was then — smooth and svelte. Maiden belly.

Then life quickened in me. And my belly was the universe, home to creation itself. A tiny bean-sized human was forming.

I bought a gold lamé jumpsuit at a secondhand store. I thought it was big enough to fit me forever. It was what I imagined maternity wear would be: an enormous flopping piece of fabric with leg holes. I have never fit into it since the first trimester of my first child.

My belly is a planet — from which two of my favorite people ever have emerged. It is a magical sphere and an arcane, analog delivery of life. Weird and wonderful is the belly. I painted it like a watermelon. Mother belly.

In labor, I was fearful that I could not get through the birthing — and I was right. It was not me alone; rather, I needed to welcome the storm of life that has blown through every mammal-mama ever. I was a bear. Bears' bellies know.

I placed a tiny being on my belly, twice ...

Over the decades, my belly puffed and muscled. It was sex, and it was domestication. It was wild, and it was soft. A muffin and a six-pack.

In my mid-fifties, I met my belly again. This time, it came calling me to my place as a señora. Blazing with worth, with earth, without apology, my belly held history. My belly said, "You are ready to stop hiding." Crone belly.

I remember, as a child, thinking my 85-year-old nana was pregnant. Her belly pooched. My mom's belly pooched.

Hello belly.

I will love you.

Fig. 20. Vivien Leung. (2026). [watercolor]

morale

Storms behind, storms ahead ...
Make a sail, catch the wind —

When the betrayal of friends has punctured the gourd of my soul, and I am leaking out life,

When the meanness of name-calling and the long wounds of history make me wish I were a small rodent living underground,

When the tidal wave, the earthquake, the drought, and the flood are all coming at once,

When the metallic ideas of a world high on technology make my lungs freeze,

When the cold distance of the logic of getting through a day of grids, schedules, and protocols becomes predictably claustrophobic,

When my sleep thins with worry and looping wires of remorse,

When the children's eyes grey with futures where they only see chaos, and as their mom, I cannot help them find the chroma, the blush, the glow to grow into,

When the bloodstains of several hundred years of exploitation are revealed to be not only in every aspect of my day, but also inside me:

continued ...

And the embers of verve and morale go dark in my belly,

And I cannot find an entry or an exit,

And I cannot place myself anywhere that does not make it all worse —

And I am both malnourished and overfed,

And appetite for life quits me,

And I am disgusted,
And I am disgusting.

Then the shine of rain on newly washed leaves is something like grace,

The curvature of my children's cheekbones is a marvel of creation,

The coarse calluses of my beloved's hands comfort raw isolation with warm skill and strength of knowing as only hands know.

The birdsong awakens my nervous system like moonlight flickering on water,

The immense history of intergenerational tending in a single recipe for soup is itself the medicine; a broth of ancient touch,

Then the devotion present in waking up tomorrow and feeding the dog sweetens, breaks open that gourd to spill my nectar of care recklessly,

Then all the creatures, with their fur and slime and toes and fins and fangs and whiskers and babies and illnesses ... come into a vision of adoration, fascination, wonder, and ubiquitous sacred performance.

I am living in a song of a trillion tones:

A mural covering in color,

A theatrical drama of overlapping stories leaping forward and backward in time,

A ceremony so profound its reverence is piercing,

A poem written in fruit, bacteria, soil, bark, and sun,

A magic spell, a potion, a deep secret cloaked in darkness and sparkles,

A love story gushing with longing, lust, and exploding in contact, skin-to-skin,

A sculpture of movement in differing rhythm, some in stone-time, some in dragonfly-time, resonating, shaping, forming ...

A volcanic urge bubbling from the center of life itself pulsates through me.

And I am again animate, again going, again again-ing ...

As every moment is a remembering that I will do everything.

Fig. 21. Vivien Leung. (2026). [watercolor]

give everything to life

The coming of the light.

Dusk on the winter solstice,

Rhythming in angles and light,

Water and time,

Focus and grain.

campfire potato

Every ancient culture has known that the medicine of food reaches far beyond the measurements of modern notions of nutrition, into the land, into the stories, into the ceremonies, into magic. Food and fire extend to all that has ever been. In other times, and in all lands, life happened around the fire. Food and story, birth and death, loss and gratitude. The fire held us together, fed us together.

And still, even fractured and dislocated, the spirit of a strawberry, the steam from rice, the crackle of meat on the fire all sing of what the body longs for. The babble of elders preparing for a feast, gossiping and remembering slips into their fingertips, into the sauce, into the folds of the bread.

This has always been so.

The meal makes us alive together. Even artificial ingredients cannot undo the spell of a soup potion. Frozen, ready-made meals in the microwave may not please the wishes of the farm-to-table folks, but they are still within the epochs of food's ability to create society, hold religion, confirm traditions, and make families.

The place of food as ritual in intergenerational culture-making is as old as time. Persephone ate the pomegranate and committed herself to the underworld; she let it inside her digestive process, where the underworld transformed into her and her into it. Chicken soup, holiday feasts, birthday cakes, holy prasad, all. Every meal is a ceremony, even the ones in the car. Every ritual is feeding of place, ancestry and future.

Have you ever eaten food in a dream?

You are what you eat, or so it is said. But you also eat as you are.

You eat, and you are. You are, and you eat.

A morsel passes the lips and enters the universe of taste and memories, teeth and textures. A morsel passes the lips and brings with it all that is, stitching everything alive, and everything that has been alive with everything that will be alive.

Bones are made, blood is made, noise is made, music is made, babies are made, history is made ...

A potato, pulled from the coals, fills the world with earthy, burnt wonder. A little salt sprinkled just where the skin parts, and there it is, a heaven of its own. Steam pours into the galaxy, reminding time that potatoes still have the thump of soil in their fleshiness. There is old, old information in a potato.

Fig. 22. Rachel Hentsch. (2026). [mixed media]

chewing ...

not the change you are looking for

The quest to "make change" in a moment when the survival of the human species, along with so many others, is at stake is, at once, anxious and urgent ... In every social sphere, there is a clamor of strategies and models attempting to chart a course from entrenched systems of exploitation and destruction toward ways of living that are vitalizing and regenerative. Inherent in this effort is the belief that it is possible to specify the details of better future social systems and behavior — and, within those systems, to pre-imagine and reverse-engineer a new way of life.

However well-intentioned, such strategies are tethered to a linear causal logic and a dream of control. There is a ghost of modernity contaminating not only how life is perceived, but how it is described, expressed, and even known — lurking in our communication, our epistemologies, and the institutional structures that hold them together.

However, the grip of linear causal logic and the drive for control are the foundation for decontextualized responses, which, in turn, exacerbate the perception of distinct, "mitigatable" crises and justify the compartmentalization behind much of science. Such responses often ignore that the present is never a blank canvas for action; it is already textured by centuries of accumulated happenings, carried into the many inheritances, biological, cultural, ecological, technological, and more, that are continuing to thread into every moment. What appears as a singular action is already the movement of countless prior actions, so that the "doer" is never acting alone, but within a momentum long in motion.

While there is an understandable comfort in the seeming practicality of these responses, especially in times of violent uncertainty, that comfort is a temporary anesthetic, masking the pain needed to stretch perception into the living complexity of how life moves.

What if the change being looked for looks nothing like the change being looked for?

What if it is impossible to correlate the actions with the outcomes?

And how shall this realm of possibilities — not yet seeable, or sayable — be brought into conversation?

Fig. 23. Nora Bateson. (2026). [watercolor]

covering for each other

Covering for each other
when we falter and crumble:

In this era —
The hero is useless,
The savior is hubris,
The leader is delusional.

And now people need each other.

Covering for each other
is when lots of pollinators
and lots of blossoms
are there to keep the meadow meadow-ing
through flood and drought and hardship ...

Not only bees, but a diversity of creatures.
Moths, spiders, flies, butterflies …

It's nth-order care.

Covering for each other
is not mine or yours,
It is in the possibility —
that we make together.

I cannot rehearse for your need,
nor can I predict it.
But I am here.

I do not know who I will need to be
or how I will need to be ...

Covering for each other
is when what you need from me
is something I do not know that
I know how to give.
But I give it anyway.

And when I meet you this way,
I find unknown versions of myself.

Preparing to cover for one another
brings us into the meadow
and out of the industry of
isolated prepackaged assistance.

Covering for each other
is improvisation in context
and healing through mutual learning ...
as we go along.

Fig. 24. Rachel Hentsch. (2026). [mixed media]

chewing ...

before the change

How does an organism know how to change?
How does it learn?
How does it evolve?
Where is the information coming from?
Where is it stored?

In my quest for learning to perceive and communicate in ways that are more in sync with the biological world, I have found that combining contextual stories is very much like the way a meadow is a conversation among its organisms. The aliveness is already and always in the multiple entanglings happening within communication — especially (for humans) in that which is left unsaid — through what the science of biosemiotics calls "multi-channel semiosis." These patterns of communication are not limited to human language — every organism takes part, sending and receiving signals across species, weaving our expressions into the wider, more-than-human semiotic fabric of life.

I began to use the term "ready-ing" to address the curious phenomenon of contextual interplay that allows for readiness for change. There is a way in which change occurs when preconditions have ripened. An example would be when someone who has wanted to quit smoking for years is, one day, ready. Or when a rift is healed with a friend. It is not even likely possible to estimate the ready-ing of these changes, but it is never possible to calculate all of the factors that have contributed to this new path opening.

From my "Aphanipoiesis" paper:

> Ready-ing is slow and sometimes fast; soft and also harsh at times; it is seen and unseen. To give it room to move is a kind of humble preparing toward undefined multiple pathways. As participants in this process, what might it mean to tend the possibilities of ready-ing so that whatever actions unfold do so less destructively?
>
> The opposite of ready-ing is to force a single-purpose outcome upon a system oriented and shaped around pathways that do not correspond to the desired change. Natural processes are never singular. Ever. To dictate an action change upon a living system that is not "ready" brings contortions of false and unrooted change, which falter, splinter, and become grotesque. An unready change is a violence in so much as it may contort the process of learning into place before the necessary undergrowth has taken place. This may lead to insidious difficulties later on.
>
> For instance, one cannot make a rule that people "must respect each other" and expect that the respect will go beyond vocabulary. Respect cannot be mandated; it is produced in the specifics of context, in unique relationships, and in the un-modelable complexity of each person's experience. False respect may be put into visible practice as rules to abide by, but mutual disdain may still infect the relationships in unspoken ways. For example, telling someone to stop being depressed, to love another person, or to be trustful is likely to produce odd configurations of the suggested changes.
>
> The unready change is a short-circuit of the necessary transcontextual mutual learning in ways that often lead to deeper insidious difficulties later on. The "goal" of a linear strategy may be to generate respect, optimism, or trust—but

> the culture of strategic productivity is ill-equipped to meet the wildness of life. To produce the goal without the ready-ing undermines the depth at which respect, trust, and optimism are the consequences of relational possibility, not mental states to adopt.
>
> Suppose one sees the world through the lens of strategy. In that case, it will be challenging to produce the conditions for respect, trust, or optimism to occur because these characteristics must be natural in their arrival: untamed, unscripted, and allowed to shape as is possible in the particularities between people (Bateson, 2023, p. 170).

I have offered the term ready-ing to eschew some of the concreteness in more mathematical language. As much as I enjoy the mathematics of complexity and chaos, I am also aware of the frustrating tendency for this tonality of description to imply a utilitarian engagement. The math itself is wide open, but the culture it lands in is not. Often, the quest for this math of entanglement becomes a quest for how to manage and direct it. My hope is that the notion of ready-ing, at least for a while, can tickle that itch to immediately implement pragmatic deeds, and allow for a submerging nourishing of possibility to take place — allowing the systems to move as they need to, without forcing preferred change upon them.

Fig. 25. Nora Bateson. (2026). [watercolor]

stories from questions

Stories are alive. They enter the living. Your story changes my story. There is no recipe for a story. No formula, no specific instructions on what gives a story its entrance into the underworld of our beings.

Stories are the weft of slippery threads pulling through the warp of truths and illusions, always iterating, pinging, luring the human soul into a re-telling. I have noticed that stories are never alone; they thread in clusters, from many directions to create a notion of reality that is coherent across generations, contexts, and reflected through experience.

Emotion, intellect, body, spirit, ancestral, and future generations are all met in story. No aspect of the living human being is left out of these undulating rhythms ... stories saturate.

For years, I have wondered if perhaps the caper to find the new story is a ruse, a mirage, seducing a culture of control into yet another self-reflexive trap. What if the story is not the guiding undercurrent of culture, but rather the boat that rocks and tosses atop it? What if the story is not the structure of the illusion, but the paint on its walls?

What if ... and here is my invitation to you, what if the narratives of culture are tapestry-ing responses to those big questions in life that cannot be answered? Who am I? Where did we come from, what are we here for, what comes after death, what is success?

What is it to be a good person? What is God? Those big questions through which a thousand stories have been weaving from a thousand cultures, since ever, can be seen as scaffoldings where our stories wriggle and tangle into tight epistemological knots.

These are the myths, the tales, the religions, the deep strata through which the stories of societies have threaded. Stories need gaps. Stories are true, and untrue. Stories must be told and remain untold. Stories, if they are worthwhile, change as they stay the same.

What is love? What is the point of life? What is it to be lovable? These questions form the premises of the stories that flow through them in the form of bigger stories. "How do I become lovable?" This question, for example, could host hallucinations of multiple narratives. It may seem very real to someone that to be loved, they must have financial wealth; to get wealth, they must win at the game of capitalism; capitalism is competitive, this is projected on nature and justifies Machiavellian strategy for wealth and power. Or the same question might hold an ethical narrative thread. Entangled with the wealth narrative, I might hold that to be loved I must be a good person, to be a good person I must know that my deeds are correct, to know they are correct I must have relevance in a doctrine of ethics and morals, I must excel in spiritual development, my form of spiritual development is righter than yours. As the two wrap each other — what could go wrong? Now, add many more contexts of narratives to this question, like politics, health, or education, and watch the complexity bloom.

The call to change the financial narrative, or the spiritual narrative, or the cultural identity of the individual narrative may be housed in another approach to narrative itself. I have found that the questions, "Who can I be when I am with you," and "Who can you be when you are with me?" invite a whole new world of story into the way we can be together. In contrast, the questions, "Who am I? Who are you?" Or "How do I become a better person?" reconfirm the isolation of the individual, thus obscuring the ecology of the many ways we may find to be alive together. Can you feel the difference?

Narratives are not causes — they are consequences upon which so much has been built that they now appear as causal. With time, culture forms into these big questions through an ecology of stories, all intertwining and interdepending upon the underlying questions. What are the new questions that will give rise to new stories? If you want new stories, don't bother changing the old ones, as you will only end up rearranging them through the same questions.

Storying into unanswerable questions is an infinite invitation, one that summons an existential need to mull together over that which will always elude us. The mulling is our wine, our nectar of finding a way through life's adventures. The simple act of telling our stories as we ponder questions that have no endpoint becomes a generator of intimate inquiry, opening arteries of perception between us. Our filters are moved by the insights and angles of someone else's stories. What was relevant in my story is different from what is relevant in yours. Your story opens gates of my perception I did not even know existed, let alone that they were closed.

And yet, to wallow around in questions that do not produce endpoints, with stories that are always changing, appears to be a useless endeavor in a time of urgency. Perhaps this is the most important thing we can do, with no agenda, no expectations, no outcomes in mind. We swirl and twist and get lost in eternal questions.This is a birthright.

If perchance, narratives of culture are transcontextual responses to questions, then the question I would like to offer you is this: What are the questions through which new stories will burst into being? Be careful not to instrumentalize this project again into a mechanism of finding the perfect one-size-fits-all story to implement globally.

Perhaps, instead of "changing the story" — ferment questions that entice meadows of stories, oceans of stories, families of stories that invite sacred vitality.

rooster

Root time, continuing something long since.
    Calls across the temporal into the dirt of before
        Ancestors and chickens go way back.

Our species is in a minuet, with the foxes too ...
    digging and poop,
        care and kill,
            basic and cosmic,
                day and night,
staying grounded in the rhythms of dawn.

And sometimes at three in the morning
    when it is still dark —
        why?

The rooster is calling for futures and pasts
    to stand on two feet with three claws,
        head back, throat open,

Oooh uh ooh oooh.

The rooster's call, beyond language, is translated
differently in every culture —
lasting lifetime span spell,
temporal chronos,
breakfast, breakfast, breakfast.

Oooh oh ooooh.

Reminding ...
You do not come from now.
You are not going toward now.
Now, nooooow, now noooooow nooooow —
is what will and what has come.

Feed us as we feed you.

Fig. 26. Rachel Hentsch. (2026). [watercolor]

infinitizing

This is not a recipe, it's a response
This is not a method, it's a metaphor
This is not a formula, it's fermentation
This is not a stand, it's a stance
This is not a rule, it's a rhythm
This is not a prescription, it's a perception
This is not a technology, it's a tone
This is not an itinerary, it's an impression
This is not a script, it's a synesthesia
This is not a curriculum, it's a characteristic
This is not a goal, it's a gesture
This is not a path, it's a presence
This is not control, it's context
This is not a frame, it's a feast
This is not a solution, it's a stratosphere
This is not a strategy, it's a story
This is not an agenda, it's an aura
This is not a meme, it's a mom
This is not a post-it, it's a posture.

Fig. 27. Rachel Hentsch. (2026). [mixed media]

chewing ...

paradox of agency

Do something! The world is on fire!

The "doing" is urgent now. And yet, when the doing is informed by misaligned assumptions, inherited habits of perception, and systems of authorization and funding that demand actions which are coherent within old frameworks of fragmented solutions and one-size-fits-all definitions ... doing is often worse than not doing.

Not doing is also doing. The act of holding back action is in itself an action, equally distributed and recursive as any other form of action. The term "agency" is often used to describe the capacity to act. Common in the English language and often difficult to translate, it refers to having a sense of control, the ability to respond, an intentionality of response, and even a responsibility to respond by making choices. It implies that there are actions an entity can take to affect itself and its environment.

There is also a dire warning against the potential of a nihilistic loss of agency: that human beings, or other organisms, upon noticing the wily movement of interdependent processes, may feel helpless and hopeless without a sense of being able to "do something" — and that this hopelessness could be more dangerous than any false ideation of control. Another territory of communication is needed to meet this issue.

There is a paradox of agency.

The paradoxes within the problem of doing and not doing are not new —

philosophers and religious traditions have wrestled with them for thousands of years. Yet, right now, the stakes are deadly in their acuteness. There is no more time to keep doing as we have done, and no way to authorize that which has not been done before. Morale is needed to keep enough fire in the belly to get out of bed tomorrow and meet a seemingly doomed stuckness in an unchangeable world.

And who is the doer? The terms agent and agency contain the notion of an isolated entity that acts, decides, and is responsible for its actions. Yet no organism or ecosystem or person or family or society exists outside the intergenerational forms of interdependent relationships with the other organisms that embody their ecosystem or society.

Each is wrapped into complex expressions of past and future that are not easily noticed (because they are implicit), and even less easily changed, because they are already anticipated. It is therefore difficult to discuss the doer without either negating the uniqueness of each living organism or negating the influences and shapings of context.

Is the agent making an independent choice? Or have the choices that appear before the agent been generated from within the contexts?

There is nothing to be gained in our learning by choosing either side of this paradox. Instead, we might learn to hold the paradox of agency open, to sense the tension between organism and context — thereby offering brief but viable glimpses of otherwise imperceptible portals of change.

While "agency" remains the accepted term to describe the doing and the doer, it is worth keeping a continual language adjustment in play — if only to remind us that it is never possible to completely nail down the edges of either.

This may seem to vaporize activation and responsibility. Yet, this observation and caution of language invites an animating experience of contextual connectedness that enlivens perception and disturbs assumptions. Such disruption is essential in drawing attention to cognitive bias, unquestionable traditions, and stuckness. Slapping a fix-it on the surface for this or that issue rarely moves the substrate of deeply unperceived habits.

unseen

Underground — is where the work gets done.

For decades, this has been true.

Healing soups made where the eyes of the institutions cannot see.

Sitting with the dying, along with waiting for the babies to be born,

This is a long work, hands must be held, sheets must be cleaned;

This is the altar of our arrivals and departures,

Along with feeding hungry stray teenagers at midnight.

Invisible toils given without notice.

The boss does not cook for your family when you are sick, remember that.

The unseen are left out of the ledger,

There is no department of supporting the support, caring for the care-givers,

Wiping the tears of the broken-hearted is honest work, best not sent to the degreed professionals.

We need each other, and we need for that need to be intrinsic.

And the moment to get to work has been here for a long time …

a long time.

I think it's time to go underground.

Stealth care ...

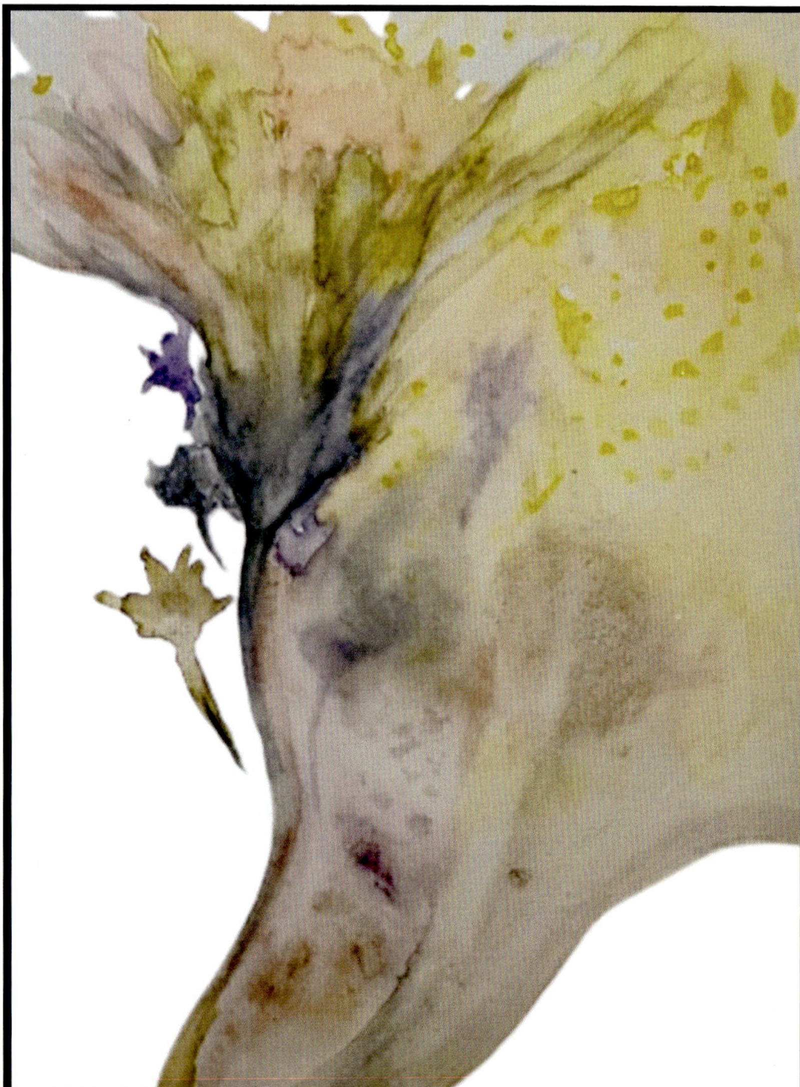

Fig. 28. Vivien Leung. (2026). [watercolor]

how much?

How much salt?

The amount that opens the flavors.

How much heat?

The amount that opens the textures.

How much time?

The amount that opens the elements.

How much oil?

The amount that opens the juiciness.

How much spice?

The amount that invites the ancestors.

gastronomical conversations

Who is the eater?

Fried words, crispy and singed on the edges.

Recommended serving for words that have lost their freshness.

Salted and hot, a wedge of three syllables leaves fingers greasy,

Dipped in meaning sauce, bright, tangy, juicy.

But this is fast food, not so nutritious.

The words are dry.... They need a lot of sauce.

Have another idiom, a handful of metaphors, no one can resist.

The baked clusters of expressions are also a delight, especially when seared before placing in the oven, they can stay raw in the middle and crispy at the cuts.

A script is a familiar recipe. A standby for moments when there is no room for risk or sparkle.

Salty salutations and minty mumblings go nicely with a tequila,

Speculations medium-rare juice with minerals and nutrients,

Where spicy side comments can ruin the evening,

But a story can mend the burnt words,

Brown them like butter and sugar,

Back to the palette, of the palate.

The crunch of a cluster of newsy bits,

The squish of a pleasantry.

The tangy zing of a just right joke.

The conversation can be satisfying.

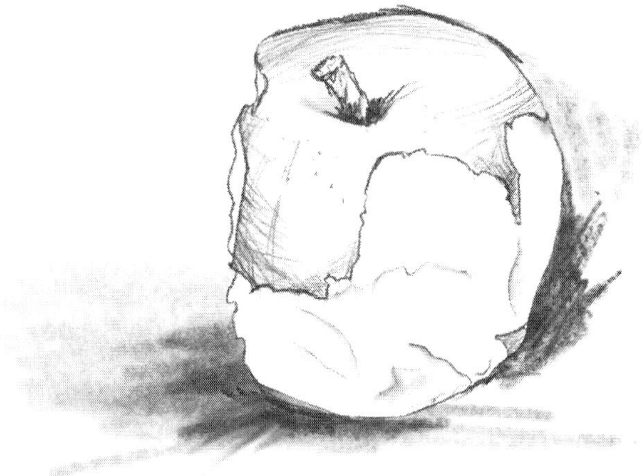

Fig. 29. Rachel Hentsch. (2026). [mixed media]

chewing ...

this is not a pipe

The "language of complexity" is not the complexity.

The "language of ecology" is not the ecology.

For thousands of years, human beings have been attempting to illustrate the nature of the universe. Philosophies and religions have depicted life and death and the sacred through the mandala, the labyrinth, the spiral, the Tibetan thangka, the medicine wheel, intertwining serpents, the endless knot, and countless other forms — symbols of the overlapping, inward-and-outward-moving, iterating characteristics of the seasons, the cosmos, phases of life, and all of these at once.

The attempt to communicate complexity and ecology through the English language is an extension of this human urge to recognize and express being alive. In prose, this communication can feel like a clumsy grasping for an increasingly nuanced perception of living systems and squishing them into language.

Attempting this is to constantly dance with a double bind — if the unnamable movement of systemic relational shifts is named, it will fall prey to being instrumentalized by the culture looking for those forms of action that are familiar and even "practical," but if the communication offered remains unnamed, it remains untended and invisible. Any semantic move beyond the tone and form of recognizable description will be filtered as incoherent by existing institutional credibility. The attempt becomes a frustrating act of surrealistic navel gazing: too abstract to matter; too slippery to hold. The communication is expected to be delivered syllogistically, and thus is unrecognizable when offered ecologically.

I find myself on all sides of this treachery — wanting, above all, to help bring life, life-ing, into language — yet wary of burdening any new perception with the baggage that "acceptable" descriptions inevitably carry to those habituated into familiar communication patterns. The form and tone of new communication processes appear disorganized and irrelevant.

How does one describe life without killing it?

> In fact, the problem of how to transmit our ecological reasoning to those whom we wish to influence in what seems to us to be an ecologically "good" direction is itself an ecological problem. We are not outside the ecology for which we plan — we are always and inevitably a part of it.
>
> Herein lies the charm and the terror of ecology — that the ideas of this science are irreversibly becoming a part of our own ecosocial system. (Bateson, 1972, p. 512)

I have been struggling with the communication around this issue — the issue of life being so inter-steeped that one cannot make tangible the future pathways of evolution or learning — throughout my adult life, and perhaps before.

My father, Gregory Bateson, and his father, William Bateson, also struggled with this issue. As a child, my father introduced me to the world I would inhabit with a double description: to invite at least two, if not more, contextually distinct descriptions to increase information through contrast. I learned early to habitualize articulating description in multiple contexts.

The breakfast is not just breakfast, it is culture, economy, history, health, ecology. But do not be tricked into thinking that double-description is just inserting multiple perspectives. The doubling is not only in adding different ways of describing breakfast, but in the ways these contexts inform each other. The culture is shifting through history, the family is shifting in relation to economy, health is changing in relation to ecological factors. Breakfast is cornflakes, breakfast is pork, breakfast is baked beans, breakfast is porridge.

The parallax of double-description is this practice of inquiry that looks for where the movement between contexts is taking place, which has always been a hard sell to those seeking narrow, static definitions. I like to sense into the way contexts are combining as though they were musical notes and chords coming together, to get a sense of the merging, each giving resonance to the others.

The first version of this double-description could be understood as the conventional epistemology that would allow me to be considered somewhat intelligible — and possibly be regarded as "sane" — within a culture that communicates and understands relationships through falsely isolated, separated labels and measurements. A second version of the double description was tacit, more difficult to parse, because the approach my father took to teach me about the interdependencies of life was not in descriptions "about" life; it was already intrinsically held within the household.

In that way, as I grew up, I added the learnings of my life into the basic premises of communicating and relating, at the core of how life makes life. This credo was a place to reconcile experiences of not only school, but the often-harsh lessons of the heart and soul.

I have made it my passion to share some of the rigorous tenderness I learned as a child with whomever I can. Yet, it is not easy to transmit this intangible swirl of vitality that underlies not only the well-being of an ocean, but of a family, or a society.

I have learned over the years that this transmission is not deliverable in direct communication. And even if delivered indirectly, language can grow into a grooved script that becomes ossified, losing its aliveness.

The communication of this living movement of learning among organisms is an ongoing practice of trying on language, finding its limits, and then trying again with new language.

Fig. 30. Noah Qwarfordt. (2026). [acrylic]

Fig. 31. Nora Bateson. (2025). [watercolor]

giving what was taken

Woman: barely so at 27, but with my newborn daughter at my breast, I became food. I became needed. Everyone is needed, and yet there is a vein of nourishment that is held in mothers, that holds mothers, and that mothers have in their hands. Each day I was emptied and filled so that I could be emptied again, like a mythological feast that magically replenishes. My godmother's left hand on my bare shoulder, her right hand on my daughter's tiny head, tipping it up just a familiar fraction to where she would latch on and begin to suck. This tiny mouth activated into pure lust for life. I knew then that she was here to stay. The mother food. "Lamb stew," my godmother said, "You need to rebuild your strength — you are still feeding two."

Transaction is not a factor in parents getting up early and making breakfast for their children. This is an intergenerational generosity, an inter-generosity. As such, the endless tending can be a burden and blessing simultaneously. Laundry, clean up, sleepless nights, huge expense, constant noise; there is no need to choose any particular aspect of the experience, they are all in the same spice drawer. The giving is constant. You give so that you give, you will give more, so you give more, you give all, give in, give up, and still there is giving to do. My mother lived 95 years, she died in my arms, and still, I feel her gifting me. I am not sure this ever stops.

A river, in the shape of an infinity sign, is the flow of generations coming and going, seasoning coming and going, dishes-forever-dishes, pots and sizzles and bubbles and baking. Burning salt, boiling the bad habits of last generations, simmering through another day.

There is nothing to do but offer as much beauty as one can to each morning's chaos of getting the kids to school. There is nothing to do but make school lunches that are as loving and delightful as possible so that halfway through the brutality of a school day, my child can feel that little box of love. There is nothing to do but to move through the obstacles and storms of a day and nudge each moment toward something that breaks away from the notion of isolation.

I suppose this could be generosity, only, it isn't ... and yet it is. This is the sort of generosity that is in play when the heart sends oxygen to the toes, not the sort that says if I have two apples, I will share one with you. Altruism can be leveraged. Oxygen cannot. Interdependency generates life by moving nourishment through the entire system, not just preferred aspects of the system. Graciousness is not just for show; it is how life offers itself towards more life. The hands that make, the inspiration that arrives, the flair of a little expression of joie de vivre may be hard to muster when the horrors of the last several hundred years seem to envelop all goodness, but when someone puts something in their mouth that is made by your hands, and they take in your ideas, your love, your spectrum of flavors, your humor, your sorrow, your stories, your ancestors, your medicine, your care ... this is an intimate moment of contact and communication.

To tend to the intimacy of this communion between us is, for me, a moment to enact a radical and ancient rebellion against all things that tell me to hold back my affection for life. I was an eighties' child, a punk rock kid. I learned to love the bleeding edge of screaming music that scratched the sterile glass of capitalistic logic. I don't need much to make something lovely for you; I don't need riches and posh ingredients. A little of this and that, whatever is lying around will be fine. The art is in the touch of the maker, not the fancy stuff.

What is true of food is true of much more than food. This is a way of life that extends to everything from how you make breakfast to how you walk, how you love, how you move through the world.

People who are not accustomed to the offering will sometimes think that they can take advantage of such gifting. They may become demanding of more. They may betray your love. I have certainly experienced a lifetime of unreciprocated generosity, but ... did I mention I was a punk rock kid? I have learned that these are the scars of generations of wounds. As these wounds heal, they itch, they sting, they become inflamed and spew rot. It can be very painful to experience what one has never been able to give or receive. These are our times. We heal each other, we hurt each other. Each moment is countless intercepting, assembling, combining histories arriving to find out if there are any new possibilities.

Sometimes, a cup of tea, offered, is a whole universe of possibilities. A sliced fruit, a cup of broth, a summer tomato with a little salt — these can be small portals to one another's emancipation from the cage of modernity. To give something, for no other reason than offering an extension of vitality: no agenda, no payback, no value added, no transaction, no ROI, no ulterior motive, no manipulation ... It's a wonderful "fuck you" to the gripping, hoarding, exploitation that justifies itself by telling us all that this is how life "is." Nonsense, this is not how life "is" — it is one depraved version that is destructive beyond compare.

Life "is" that moment when a baby's small mouth latches onto the nipple, and a continuum of physiological, cultural, intellectual, communicational, emotional, ecological, bacterial, immunological, nutritional, hormonal, intergenerational absolutes are met. The bottle works too, nestled in the arms of the parent who cares, the transmission is there. These moments are not relegated to babies; they are there when we care for the sick, care for our elders, make coffee for our lovers, bring snacks on a journey, give our pets treats, water the garden. No hold backs. There are many ways to be alive together.

rain

The machinations of confused and hurting human beings can be a lot in these times.

But there is rain.

***

Raining resonances, ring another honesty.

Washes of wind with wooshes of water,

Rippled rhythms — a carpeted stanza of droplets,

Yellowing leaves sway into autumn.

Sky pulling into soil,

Soil breathing into everything.

Nude in cool showered shine,

A pause — sparse plops contour

the quiet of new mud.

All branches are listening in,

Punctuated rest, stopped stillness.

Mulchy

musty

wetted

dusty ...

Sprinkling

pickling

soggied

saplings ...

Organisms tasting time.

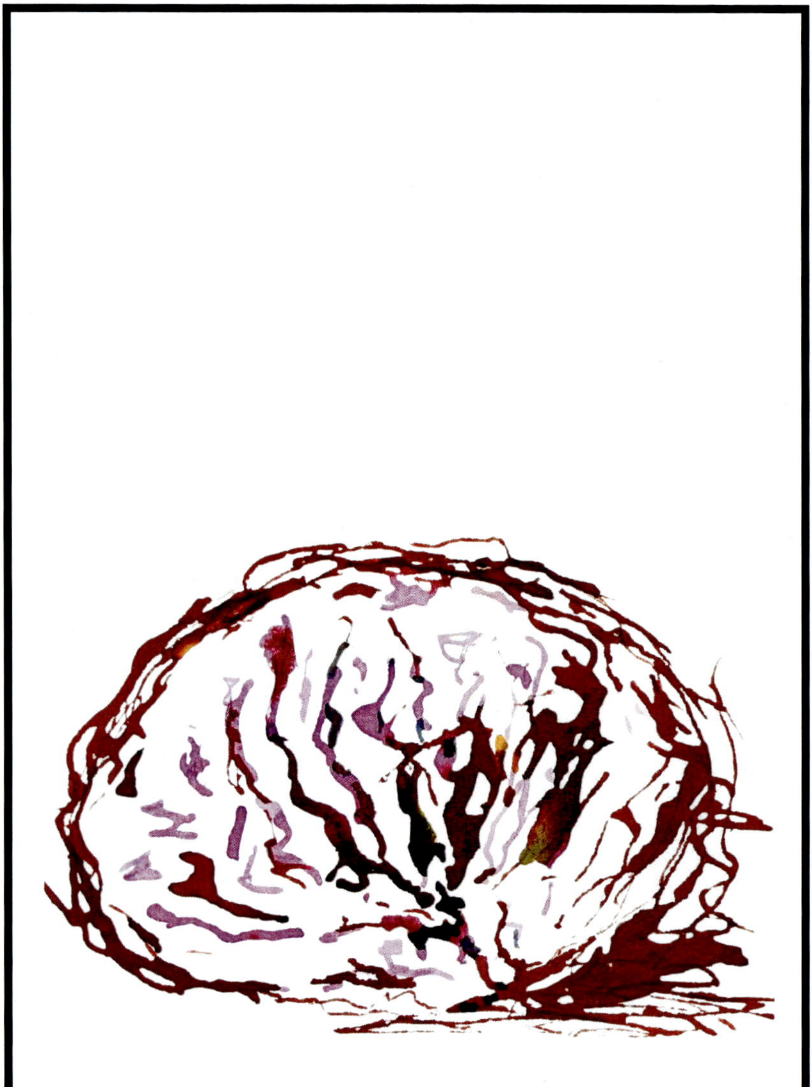

Fig. 32. Vivien Leung. (2026). [watercolor]

all the life forms

You cannot stop being wonderful.
Greetings to your gusto.

Hello, bursting communion!
Aliveness unfettered — I see you
Blooming
Unquiet

No matter the weight. It's easy lifting ...
Coloring up the grays of impossibility —
With excited vitality,
In membership of life.

Uttering vibrancy in rhythms.
Giving more than everything,
In a kiss, a breath, a glance

Fiercely fantastic!
We are among the tones of life itself.
Resounding, releasing, recognizing each other.

meatballs by stepmother

What was I thinking.

This is not a question; it is a cringe.

My first night alone with my Swedish stepsons. They were 10 and 7. My husband was at a gig, and I was going to cook dinner for the boys. I was nervous. I felt a horrible competition with their mother.

Why did I feel that way? I cannot say. Too many generations of women being measured against their recipes — I suppose. The kids also needed to be loyal to their mama's food.

Step-parenting is complicated. Meanwhile, notwithstanding the intergenerational, intercultural shit-show of mutual violence, I needed to feed them something they would eat. They were picky.

I could not have screwed up more thoroughly ...

I asked them what they wanted, and they tried to pick something safe. Bless them, they knew too that this was a marker moment. They asked for meatballs, thinking it was a basic and easy request. Little did they know. I had no clue how to make Swedish meatballs. I needed to give them what they asked for. How could I not?

Forgodssakes, when in Sweden, never make meatballs for Swedes, unless you are Swedish ... which I am not. I called my husband in a panic, asking how to make Swedish meatballs. What a silly question.

First, you need to have a Swedish grandma. Then you need to attend several generations of festivals in which meatballs are the cornerstone of the holiday feast. Then ... you need to actually like meatballs, which I do not.

The fiasco was epic. I still recoil at the memory of the boys pushing around little blackened blobs in the grease on their plates. They tried to try. There was no way they were eating those nasty little oily orbs I produced.

My phone rang.

It was a good friend, an elder Swede ... who set me straight.

I answered the phone in tears. He asked what was wrong. I told him.

"You do not have a Swedish mom or grandmother, you have no business making meatballs for Swedes, they will know you cannot fake this ... go to the mini-mart and buy the readymade meatballs, don't be an idiot."

That ... is a friend. He was right. I have made my stepsons a great deal of food since that time — they have loved it all — but never, ever make meatballs for Swedes unless you are Swedish.

lawnmower world

Attending to vitality requires morale, and also gives it.
Each day is impossible. Each day is in service to possibility.
I am a meadow, with you, blooming and pollinating,
— living in a world of lawnmowers.

Knowing —
The clanking culture of destruction like my own hand ... in my own hands, with my own hands. The same hands that heal and hold babies.

Knowing also that life outlives un-life. Tiny, fragile, determined little green things all wiggling around being pulled into life, pushing into life.

Together:
They are in the darkness, lost, and on their way. Each new organism is ancient, arriving in a continuum. A million years, a hundred trillion family members. A single moment of all living things is infinite generations of backward and forward.

Happening.
Right under my feet, right into my breakfast, right there, right here!

Meadowing,
Tending the delicate interplay, the deep time, the moment, the possibility of ongoing worlds of: bacteria, insects, plants, fungi, rodents, birds, reptiles, and mammals — me.

Learning ...
Doing, in fractal musicality ... a composing of forms. Such doing is allowing for the doing that moves among,

Among-ing us in creatura.

The lawnmowers will come, with relentless appetites. The whirling blades of productivity and efficiency. And the meadow stands no chance. They are fast and cold, sharp and brutal. They do not have the silence of either predator or prey.

Smashing metallic things do not worry, or cry, or scar, or fall in love. They are unfeeling, unknowing, un-sensing, un-alive. Then, yes, those other guys, the dandelions, the horsetail,

The punkrockers.

Unstopped by the carousel of twisting knives. I am learning grace from them. Garages fill with poisons that promise to knock them out.

Ha. It is a fool's task to un-life the world that gives us life.

Horrible hubris.

With an infection in our connection. Meadows still tickle themselves with fluttering, scurrying, slithering organisms.

Slime and rot and stink,

Mess and mistakes, and muddles ... these are the messages of morale — to fiber the souls who meet another day in the lawnmower world.

Fig. 33. Noah Qwarfordt. (2026). [acrylic]

invisible ingredients

When my mother would cook, she would always dote on the veggies and meats. "Oooh, such a nice chickie." She would coo to the chicken she was about to roast. "Such lovely little potatoes, nice green lettuce ... " And so it was, she loved-up each ingredient vocally if someone was in the room, and if no one was in the room, I am not sure what she did. She broke from the trends of her time and breastfed me.

She never once said she loved to cook. I do. I love to cook. But then, I am of a generation that was given the option. She was not. She was an adventurer, bound for strange lands and people with wild ideas. She did not want to be stuck in the 1950s housewife role. The best way not to get stuck was to be very good at it and then offer those skills on her own terms. She could serve up a meal for 10 without blinking. I hear that she could also bake. But all that was before I was born.

One day, when I came home from 4th grade, my mother made a declaration that she was all finished with the business of meal-making. No more. I panicked. But no one else seemed to be concerned much. It was as though they knew she would not actually stop cooking. Or, perhaps they just could not imagine a world without the food magically appearing at the table each night.

They were sort of right. But she was tough. She did mean it. She stopped cooking for about ten years. We moved to an intentional community, and there were communal meals. She became a health nut, growing wheatgrass in the house, eating raw foods.

My dad had been diagnosed with cancer. Learning about health food was her way of tending to him. He, an Englishman, brought up on meat and potatoes, mocked the health food, but he was also grateful that she was learning about something. I suppose she was also in preparation for his death, which meant she needed to do whatever she could to take care of herself. I understand now that she was in menopause, she was caregiving, and she was a woman born in 1929 who was soon going to be flying solo in a world she did not feel comfortable navigating. She was proud. She never said any of this. She acted like studying nutrition was the calling of her lifetime. Perhaps it was.

Englishmen like my father are no strangers to the kitchen. He liked to eat, and he liked to cook. He made fried everything. Fried bread, bacon, eggs, tomatoes. As a child, I thought this was all wonderful. The sourdough was soaked through with the grease from the big cast iron pan and then sizzled until golden brown. My father's cancer was in his lungs; despite this diet, he didn't have any heart issues. Ironically, it was my mother who had the heart troubles.

Once, I was going into town with some friends, which is an hour and a half drive from where we lived in community. My father, who was quite sick by this time and very sick of the community menu, asked me to get him some lamb chops. I was ten.

Having been around so much vegetarian fervor, I did not know that lamb chops and pork chops were not the same thing. I bought pork chops. He was really looking forward to those lamb chops, but when I gave him the pork chops, I could see his delight crash. "Oh well," he said and brushed it off so I would not feel so bad. But I still do wish I had delivered him those lamb chops.

Any food we had at home, except for some of the 1970s early health food recipes, was delicious. The bland tofu dinners were not my mom's fault; she was hoping to help her husband heal, and herself as well. But there was really no hope; the early health food was dreadful. Vegetarian food has come a very long way since those days, thank goodness. If you know, you know ... carob cakes.

And even though I loved that fried bread, it was my mother who taught me to cook. The knack of it, the gist of it, the gait and grace ... it all came from her. Her touch. I was able to build upon her repertoire, but that was because she gave me such good foundations.

I never once saw my mother or father use a cookbook. They just put some of this and that in the pan, tasted often, and it was always good. Food has always been family, health, conversation, culture, seasons, nutrition, and punctuation for the beginning and end of each day.

Food is the way we love.

labyrinth

It looks like a brain.

And surely the movement of understanding and the arrival of ideas form a rhythm of always brushing, caressing, crashing into other ideas. They remind each other of each other. They are familial, they argue, and they mourn. While getting somewhere, they go nowhere — iterating. A wild thing lurks. Captive, under the scaffolding of culture, our dreaded and yet beloved creature has ideas of its own to show us. To meet this creature requires both entry and exit from existence as it is now known. The beast roams at the center, beckoning unpredictable visitors. Tender afraid-ness side by side with ferocity, confident and terrified, luring us into untamed learning.

It looks like intestines.

The nourishment crosses membranes, filtering into metabolisms and circulations. The guts are a squishy winding-ness of tubes of tissue that share information and transform onions into conversation. The tubular hallways hold one another in place as they sidle up and whisper proteins, sing minerals. These bacterial dance halls dizzy and intoxicate a gala of organisms. The dance is imperative, and travels each being to each other being, for a moment of contact. This is the allowance of evolution. To commune with many is to gather and offer possibility.

It looks like the fossilized grooves in sandstone formed by ancient worms.

Dignity and humility in elegant symbols of something disgusting, that is as beautiful as life itself. You can put your pinky finger in the concave cylindrical avenues, and remember, try to remember your slime. You once were a worm. Worms remember what came before worms. They know what you are hiding. They show you.

It folds the way a day is touching tomorrow and yesterday.

And in this chronic confusion of trying to flatten things into linear time, linear cause, linear solution, linear plans, linear dreams ... the lines are bound to each other, licking and nudging the knowing that seeps through the tubes. Don't worry, next year is already ancient. Frustration of wasted time is just the future bossing around the past. These false binaries of timelines are children bickering in the back seat.

It folds the way the generations fold, old and young, stories overlapping.

Here, our scars and wounds sing, healing through grandmothers, through new acorns on moist forest floors, through algae housing translucent eggs. We are each other. This can be longing and merging simultaneously. We are nothing but squashed spirals remembering the dangers of symmetry. Staying wobbled, we find new pockets of old, fermenting inspiration. Other people's stories are threads that help us find our way.

We are finding a way.

We are losing our way.

It folds the way that I see you becoming and be-going.

Living requires dying. Recalling being alive is as blurry as recalling death. Continuing and discontinuing. The struggles give comfort, reminding us that there are walls to bang into. We are not random. We are not planned.

It holds secrets, but with no closed doors.

The corridor is open, and it meanders. There is no trickery, no manipulation. There are no lies, except those that have found their way into our habits.

The curvatures keep coming and going until the coming is the going and the going is the coming. Anxiety is unraveled. Let's become unraveled.

It holds the lostness but with only one path.

One might imagine that a single road should be hard to get lost on. Like chaos math that does not measure or solve, the trail's irregular coiling gesture is the world, the whole universe perhaps, it is infinity. Sometimes getting lost in infinity is wandering in ecstasy.

It invites as it lures.

Returning, returning with each turn that re's: re-this-ing and re-that-ing, re-membering and re-patterning, and re-versing, re-routing, re-thinking, re-touching, re-foresting, re-living, re-jiggering. Come back, go away, come back changed. Go away, new.

It invites solitude through connection.

Nothing to hoard. Nothing to give. Nothing to shield, nothing to shine. Nothing, because everything is already vividly soaking in everything. Leaving only the generosity of impossible truth to clothe us in our nakedness.

So here we are, alone together, in the digestion of so many cultures and civilizations that have brought new life and wonder to their time, touching the living and the dying, the slime and the grace, the wiggle and the weeping, the harmonizing and becoming out of tune, into a new resonance.

There is no solving the labyrinth.

No escape.

... only the sometimes-slippery tissues of learning.

... only the sometimes-roughhewn stone containers of learning.

Fig. 34. Vivien Leung. (2026). [watercolor]

Fig. 35. Vivien Leung. (2026). [watercolor]

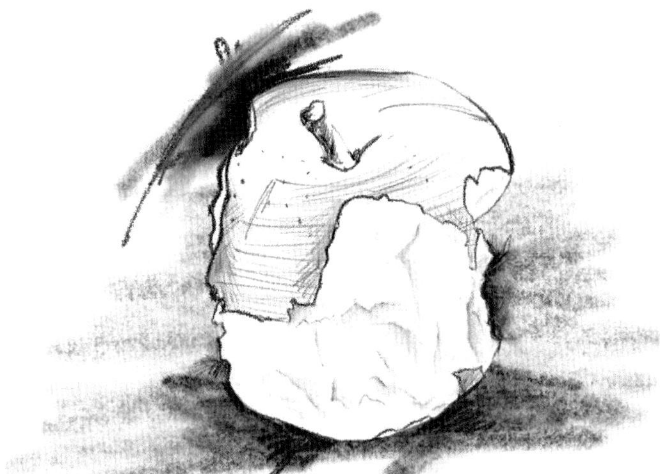

Fig. 36. Rachel Hentsch. (2026). [mixed media]

chewing …

orders of actions

As the recursiveness expands, the nth-order comes into view. Understanding the importance of the term "nth-order agency" is tantamount, I believe, to the possible survival of the human and other species. To this end, I am eager to bring the term into use. Yet, I do so reluctantly because the term itself conjures impressions of concentric circles of causality, and sadly, concentric circles are not a wild enough image to hold the swirling complexity of mutual learning in life. Strategizing several moves down the chessboard is not what is being discussed here; that is still linear, correlative change. Granted, playing out several moves ahead in one's mind requires sensitivity to the relational movements of other entities, but the nth-order is reaching into contexts that are unrecognizable and beyond the gameboard.

Loosening a linear habit of articulating causation is the issue at hand. Blurring directional assumptions of how something has come to be, or predicting through reduced context what will happen, requires a practice of sensitizing to another logic. This is a Warm Logic, a logic that does not mock the variables in play at every moment. The causal logic of infinite transcontextual processes is difficult to grok. In a meadow, every organism is changing every organism all the time. The responsiveness among the organisms is boggling to comprehend, let alone to predict. The disorientation of trying to comprehend this infinitude is essential to the possibility of reorientation to an understanding of causation that is multidirectional. The catch is that the infinite changes of changes cannot be captured in a bigger spreadsheet. The approach to inquiry itself is shifted with the impossibility of explicit definition, control, and metrics. The inconvenience of infinity is a gift of humility that reshapes the idea of research and information entirely. To think about a meadow, or the birth of a baby, or a meal, it is necessary to enter another attitude of study that finds the infinite to be imperative. Nth-order is barely enough.It is still too easy to fall into the trap of trying to control nth-order change by picturing circles within circles as layers of recursion, without noticing that this infinitude of causality is moving in all directions, as well as backward and forward in time. I am cautious of the term, yet welcoming it nonetheless.

Everything you and I do takes place at the nth-order. You cannot simply draw your line of responsibility at the edge of first-order action. It goes far beyond that. And yet, every response may look like a first-order response, even if it is shaped by considerations of the many reverberations a single action has within a complex system.

Equally, every first-order response carries out into nth-order repercussions, whether or not they were considered. If someone breaks their ankle, the need to get medical care to reset the bones may appear to be a first-order response. But both before the breakage and afterward, the person's nutrition, mood, habitat, conditions of the day, family support (or lack thereof), staircases in their living arrangement, etc., are in play. No action is solely first-order or nth-order — all are entangled.

Perhaps one of the reasons the ghost of first-order illusions is so potent is because of a shared experience of first-order strategies working marvelously. I am hungry, I eat. I have mud on my shoes, I clean them. I want to sit down, I find a chair. I am having a party, I invite people. But the illusion lies in the way my noticing of effectiveness is not only registered at first-order. Pesticides were very effective at killing insects that demolished crops. Antibiotic medications were miraculous at stopping infection. Selling products that have built-in obsolescence has sustained enormous profits for companies. But in each case, the other contexts of consequences were not considered, neither in the strategizing of finding a particular solution, nor in the sensing of other contexts that might have become outcomes.

The lack of perception of transcontextual consequences is then a cost factor in doling out responsibility and accountability for any destruction that reaches out from the perceived successful solution. The increase in cancer rates is not the responsibility of the farmers who use pesticides; the superbugs that have evolved in relation to antibiotics are not the responsibility of the pharmaceutical companies; vast amounts of garbage and unmitigated use of natural resources are not the problem of the companies whose products end up in landfills after two years of use. It is convenient not to try to factor in the "unintended consequences." The coronavirus opened up a world of digital communication that has altered the way people date, go to nightclubs, and even how, when, and where they eat. Strange, isn't it, that a virus can trigger a food delivery habit years later for kids' after-school snack time?

Today, I enjoyed a morning coffee. This is a first-order perception and description of the situation. From this description, there is an illusion that the enjoyment of the coffee was the action. Using a language of "orders" is awkward, and it is certainly confusing to try to parse the concept of these orders in terms of how they can be differentiated.

Nevertheless, the language of orders is one way to try to describe the difference between an action that is identified as being tethered to a direct effect and one that reaches beyond direct effect into more systemic ramifications from the action.

First-order action might be like changing a tire that has gone flat or flipping a light switch to turn the lights on; second-order adds a recursive action: I like to drink coffee in the morning. In second-order description, I am observing my habit and something about my character in relation to coffee.

"It's not just that and nothing more" (Bateson, 1972, p. 272).

I can go on in recursion to make the description so that it invites description of an order that is immeasurable, or nth-order. As I make my coffee in the morning, I am comforted by the routine of brewing, the lovely roasted aromas that fill the kitchen, and the warmth of the rich liquid, which soothes the throat that is dry from the night. I believe that caffeine is what gets me going in the day, and it may be, but the ceremony of stumbling around in my pajamas, making my morning mug of coffee, is also a physiological greeting of the day.

It is wrapped in history and experiences that have been whispered through mornings since before I was born. And while for me the cozy morning wake-up moment signals goodness, for many of those who have tilled the soil for growing the beans, produced the sun-protective hats, and sewn the clothing for the people who worked the fields, fed them breakfast, taken care of their children and elders, and then roasted the coffee beans, transported them, marketed them ... the story of my coffee is also a story of exploitation, soil degradation, water consumption, child labor, poverty, and international injustice.

Nonetheless, my gut churns, my digestion is stirred, my sleepy cobwebs are lifted, and I go out into my day. I do things; I work to earn the money to cover my home, food, coffee, and other consumptive habits. I have meetings, run errands, tend children, tend my aging mother, clean the house, walk the dog, have a cup of tea, make dinner. The morning coffee meets my microbiome, my nervous system. It makes me talk too much sometimes, which can be fun until my partner realizes he is late for his day.

We stress, we rush, we forget the lunch box, the homework, the milk turns from being on the counter all day. I am in an erratic dance with all these doings and non-doings that shape the next pathways. I need to buy lunch, we have to call someone to find the homework, the meeting has to be rescheduled ... the infinitude of the redirections is boggling. The adjustments ring across eras, cultures, economies, and generations, and in doing so redefine the conditions into which the possible progressions emerge.

Let us give some more specific examples of what I, at least, describe by the term nth-order. These examples show how nth-order is the way ecologies move, with all the organisms shaping themselves through communication with other organisms, ideas morphing in response to other ideas.

Gregory Bateson wrote:

> Consider a stag's antlers — the horns on his head — of course, these are to ward off mountain lions and to fight other stags. Perhaps more the latter than the former. That's more for dealing with other stags than to deal with mountain lions. That's fine, they're weapons, yes? A weapon is a fine thing: we know how to make it, and we do better than stags do, and we can kill each other, I guess. Now, the trouble is, which you see that the lady — the doe — probably thinks those are very decorative objects, so that they have, you see, two sets of purposes: fitting twice into the whole system. And everything that fits twice into the whole system becomes, in fact, a bridging point — a node — in a network tying things together, and it is things tied together that is what we're interested in. I talked yesterday about the horse and the grass. The teeth of the horse are, of course, to get stuff to put in his stomach. But the teeth are also to prevent the grass from going to seed. His hooves are for locomotion, of course, but his hooves are also to make the grass tiller. And if you begin looking at the biological world, you will find that it is over and over again tied together through double uses in this sort of way. Where everything is as the Freudians used to say, over-determined. It's determined from the side of sex life, it's determined from the side of food getting, it's determined from the side of whatever else — locomotion, whatever else that goes on. And this is one of the ways in which these larger networks in which our feelings are deeply hooked up, you know, with a fine combination of feeling between having a weapon and being admired by the opposite

> sex, you know. We even buy automobiles for that purpose. Now you've got an automobile as a degenerate weapon. I understand it's also used for transportation. So it has three uses: it's to charm the ladies, it's to move people across country, and it's to kill the other automobiles. No wonder we love them, [laughing] (Bateson, 1975).

As modern observers, limited by what is considered an ordinary education, we might fail to notice the processes of transformation that occur in nature. The appetite of the lion does not appear as though it is causally contributing to the conservation of the savannah. But when it kills an antelope, the lion is sometimes lessening the population of grass-eating animals, thereby protecting the grass, which holds the soil and keeps it from parching in the dry months. The lust of the lion's hunger is fed by the antelope, but at nth-order it is contributing to the ecology of the grassland (unless there is a severe imbalance in the ecosystem).

The tendency to place cause and effect within the bounded relationship of the lion and the antelope obscures the multiplicity of essential effects-of-effects the lion's hunger meets, keeping in mind that all the organisms in the ecosystem with the lion are also engaged in many directions of relational interdependence.

What this concept of the nth-order turns out to be about is a multidimensional transcontextualism of what we might even call story — leading in ramifying directions from an innumerable set of past antecedents to a present, which is not "just itself and nothing more."

Indeed, a single defined event will trace connections backward, forward, and laterally. We tend to think of these connections as interacting with each other to form a network, in which circular causations can be traced by observers in multiple ways.

Nature never does one thing at a time. The earthworm aerates the soil with its movement underground, allowing for plants to grow their roots deeply, and the earthworm nourishes and repairs the soil with its poop (vermicompost) — it is also food to lots of animals. Unlike you and me, who try to pack too much into the day, the earthworm is not stressed out by this multitasking, as it does not have to consciously attend to its many relational doings.

Each organism in an ecology is in communication with countless other organisms. This is life making life. Earthworms are also good for going fishing, grossing out your friends, featuring in horror scenes of corpses, or one appearing as a wise being in Dune. Mythological stories about worms and their serpentine cousins are found in nearly every tradition. Earthworms, like all organisms in ecologies, are simultaneously implicated in many forms of communication, including some that are in human relationships. So — where is the edge of the earthworm?

Several hundred years of thinking of causality in terms of industrial or mechanistic ideas have obscured perception of, and grace within, the swirl of infinite transcontextual consequences. The way each action and non-action arises within contextual pressures, and the way they extend into continuing responses to responses to responses, is unfathomable. Impossible to calculate, these ecological iterations bring important questions to familiar notions of action, and agency, and purpose.

The impossibility of reckoning with an infinitude of knock-on effects — the ripples and changes in what is sometimes called externalities — makes reductionist models of change all the more seductive. The appetite for outcomes, deliverables, and trajectories of preferred change is stored in countless binders of project reports, all of which have been well-intended but have failed to move the compass of what is considered change. The necessary redundancy of any ecology defies isolated effects.

Externalities are therefore not external to anything, except the scope of consequences drawn by someone who did not include the rest of the relationships; that scope was always an illusion. Within these networks, we trace connections through the filters of relevance that are in place for us. No one traces all of the connections. To evoke my book title, *Small Arcs of Larger Circles* (Bateson, 2016), when our goals are purposive, we tend to follow small arcs of goal-directed cause and effect, while when we let go of purposiveness, sometimes we might perceive the larger circles of ecological relationship and process. But it is not as though we can ever map these relationships completely.

The lesson of nth-order is not to build new kinds of omniscient maps or diagrams. Instead, it is to develop a humility in the face of what can never be fully known. As Roy Rappaport (1974) used to say, "knowledge can never replace respect in human dealings with ecological systems."

Fig. 37. Rachel Hentsch. (2025). [watercolor]

a lemon

With salt and olive oil, a little ground pepper. Mixed with a fork and poured over crisp lettuce.

With bubbling melted butter, in a small dish for dipping artichokes.

With egg yolk, butter, and white sugar, on a low heat, a bit of cornstarch and water, becomes a curd for bready things.

With garlic, it sometimes turns blue. Fresh bursts of juice squeezed over a piece of fish.

With tahini and ground up garbanzo beans, salt, and olive oil. Zested in a sauce, in a cake, on a pasta.

In each placement, the lemon is in relationship with the other ingredients in ways that change both the lemon and the other flavors. The lemon is just a lemon. No two are the same. The ones with the smooth skins are the juiciest, but hard to peel. As children, my friends and I would write secret, invisible messages in lemon juice that would show up when the paper was heated up. I have always used a fork to juice a lemon, and then to scoop out the seeds. I really love the yellow — the fresh sour yellow light that the lemon brings.

Getting to know the way lemons are in different combinations is a way to get to know the lemon. Getting to know myself in lots of different situations is a way to get to know myself. Getting to know any living organism in different situations reveals the way they respond to new contexts, and the way new contexts respond to them.

Have you met me in a traffic jam? Or a dinner party?

Have you met me in the soft hours of the night?

Have you met me with my kids?

Have you met me at the doctor's office?

Have you met me on the phone with the insurance people?

Have you met me when I am painting?

Have you met me at home?

Have you met me with my best friends?

Have you met me in a forest, or a city?

extemporaneous du jour

A recipe for finding stuff to eat.

This is life. Not every day is a festival. But perhaps there is something edible around. How does one pull together something delicious?

Having absolutely no idea what other people usually keep in their refrigerators, it should be far-fetched to write a recipe for what is for dinner when there is no plan.

What is possible?

What is possible is always in the details, and the details are always holding possibility. Somehow this is so. The question then becomes: will one perceive the possibilities that are awaiting?

It depends on the gaze, it depends on what matters, it depends on the mood, it depends on the company present. And of course it depends on what is in there in the refrigerator, what is in the cupboard, and what is in the backyard.

The art of spontaneous scratch cooking has everything to do with smiling into the challenge. And, it has everything to do with knowing that what you make, you make into something beautiful. No matter how simple or strange. This is the way all good things begin. Later, you might think, "how come I never tried that combination before?" — and vow to make it a standard. Like all improvisation, it takes practice. Artisans spend decades, generations, and even entire eras homing in on the nuance of the materials of their craft.

While every day is full of this extemporaneousness, one of the most ridiculous trials of this practice was a fateful Christmas dinner in Copenhagen. Our family was traveling, and we miscalculated badly. We had gone online and made reservations at a restaurant for Christmas Eve, thinking that since we were going to be in an Airbnb, it would be more fun to go out. But, it turned out that the online reservation site was faulty and had booked us a table for a restaurant that was not open on Christmas Eve. Most Danes would not have made this mistake, because almost all restaurants are closed on Christmas Eve. So, there we were, no Christmas dinner.

At first, we thought we would go to the shops and buy things to cook, but they were all closed as well. We went back to the little apartment to get out of the cold. As it turned out, there were a few leftover veggies in the refrigerator of the rental apartment: a carrot, a potato, a pepper. And, downstairs under the apartment building was a 7/11. I decided I would find something to do with those few veggies and whatever I could find at the mini mart. But the minimart had almost nothing. No fresh things. I did see they had some hot dogs on the grill and some instant rice.

That Christmas, we had the most delicious stew of veggies and hot dogs over rice. And of all the beautiful and extravagant Christmas dinners I have ever produced, that was maybe the most delicious, and certainly the most memorable.

Familiarity:

It's a matter of being acquainted in many different ways. The textures, the saltiness, the way a thing heats or chills ... to know an olive oil is to know its thickness, its spice, its urge to dominate, or disappear, to see it be changed by a lemon, or salt, to watch the way bread soaks it up, to pour it over a tomato, or into minced garlic. Ah, I know you. You are the olive oil that speaks in greens and dances with fresh basil ... cousin to the olive oil that welcomes the balsamic and mustard. That one is shy.

It's a matter of noticing the veggies and how they are together. Do those carrots turn everything too sweet when cooked? Does that celery crunch meet the leafy piles of spinach leaves in a way that is pleasing today? Perhaps those tender little potatoes would be better seared in a cast-iron pan and then slow-roasted in salt and rosemary than tossed into a soup. Yes, the carrots are too sweet. I will keep them raw, long peel them over the salad with the thick balsamic. That celery is going to have to wait. Today is not its day. Spinach leaves, lightly steamed, butter, brave olive oil, and lemon juice. Salt. Lettuce leaves with long, grated carrot ribbons. A little mustard in the shy olive oil ... Now I need some cheese or nuts, or protein for this salad.

Here is a brie, it's not opened. Here is a corner of a chunk of cheddar, here is a wedge of parmesan, a little bit of mozzarella. Mmm ... mustard, and carrots, and ... no. None of the cheeses are calling me. I keep rummaging and find a bag of walnuts. Yes, a little pan roast and speckled over the top of the carrots. Maybe I will add a squeeze of an orange to the dressing to bring a fresh, sweet kiss to the nuts.

Improvisation is not random. Resting into the way the players contrast one another and bring out different aspects of their characters is very similar to making a meal with no plan. The ingredients will tell you who they want to play with, if you know them. The meal will form in the process. The better friends you are with the ingredients, the more they can speak to you.

So much is considered. What needs to be eaten before it goes bad? What is being saved for another meal? What did we have yesterday? What is the weather like today? What am I craving, what am I bored with? Who is eating? What is the mood? Which dish shall I serve this in?

I find that I sometimes decide what to cook based on an attraction to a particular dish in the cupboard. The earthenware serving bowl is appealing, and the next thing I know, I have made some roasted roots to serve in it. The vibe is warm and grounded. Some days ask for earth and roots. Others are days when I serve right out of the pan, sizzling and messy. Informal, loving, lazy. This is good too. There is no better or worse way, just the attunement to the moment, to the context.

Improvised meals can be breakfast for dinner, or fried bananas for lunch. A quick soup or stir fry, a roasted mash or a hash. All are potentially the best meal you have ever had.

This is systems thinking in the kitchen. Responding to what is relevant in the moment, in the fridge, in the family, on that day, in that house, with a particular group of people. The combinings are more than just ingredients. They are social, seasonal, cultural, time-dependent, and so much more. This is a practice, one that can be extrapolated into all sorts of situations in life. Listening to the belly as you gaze into the fridge is akin to the way musicians on stage listen to each other and the audience.

What are the potentialities of the next combinations? What is attractive?

Fig. 38. Rachel Hentsch. (2026). [mixed media]

chewing ...

trust

> We live in a world in which distrust and greed and violence masquerade as common sense, and in which the pathways of distrust and greed and violence are rapidly becoming self-validating. By following those pathways we create the social and international structures, the premises upon which we must live. By choosing the "common sense" of distrust, we choose also the progressive truth of distrust. We cause horror to become the only pathway to wisdom. (Bateson, 1979b, p. 8)

It may look as though my father is pitching for greater capacities of "trust, generosity, and nonviolence" here. But that would be missing the deeper ecology and the systemics of what he is saying. He does not imply here that trust and generosity are better versions of common sense that should be propagated. No. The issue he is pointing to is that distrust is a second-order response to a world whose common sense is premised upon a way of understanding that is centered on individual, separated, controllable, ownable, measurable bits.

Instead of rebuilding trust, (whatever that may mean), he is suggesting defragmenting the way the world is perceived. Honoring the connective tissuing of life will beget trust. If the relationships don't matter, distrust is to be expected.

Fig. 39. Noah Qwarfordt. (2026). [watercolor]

the eighth deadly sin

Looking, for examples, moving images of blobs merging.

Listening, for examples, chords combining.

Tasting, for examples, flavors mixing.

Noticing:

Sensations blending, practicing sensitivity to the inextricable, intersteeping of life.

Calling for moments when people are ensembling toward tender, yet raucous, yet funny, yet intense, yet elegant, yet outrageous, conjoining — being alive in an alive world.

We are grapes in the wine ...

Is it the eighth deadly sin to divide those who might otherwise blob together?

Fig. 40. Vivien Leung. (2026). [watercolor]

communication is sacred

When Gregory Bateson and Margaret Mead met in 1932 in New Guinea, the world was gearing up for a war against fascism. The problems with confronting fascist ideas have not changed since then. The trap of trying to confront fascism is that it grows stronger with polarity, and the problem with not confronting fascism is that it grows stronger when it is not met with resistance.

So, what can be done? Rallying against a group that believes itself to be superior further ignites a sense of righteousness in their polarity. But without counteraction, the momentum of the hateful cause grows deeper and wider into communities, demanding more loyalty, and more exclusion.

Most attempts to stop fascism seem only to generate it. These ideas and the work that began back in the 1930s in response to Nazism are relevant today in response to the escalation of divisive algorithms currently careening societies into authoritarian tech nightmares.

A New Science

Gregory began to ask how fascism takes hold of societies and what might make it more difficult for those kinds of ideas to find purchase. He began what would be a lifetime of work addressing the thinking that was susceptible to fascist ideas and how to meet this sticky problem.

One of his observations is that when any aspect of a living system is torn from its contextual relationships, it can then be exploited. How a description is made of a person, a family, a community, a culture, or an ecosystem matters.

Does the description hold the complexity, or does the description sever the relational connections? The more relational, contextual understanding there is, the less likely polarities are to take over.

Thus began the work to create a scientific inquiry that would study how systems are entangled. These became systemic, cybernetic, complexity, and chaos theories. This new science, he hoped, would bring contextual information where reductionism had left gaping holes in the knowledge of culture, society, family, health, and ecology. All of these share a vital sphere of communication and relationship.

As living organisms, we commune. As living organisms, we perceive contextually.

But what is the communication? And what context has been perceived to commune within?

If the communication is divisive, and the context is one that favors the control of machines over the wilderness of life, the very core of membership with each other and the living world is eroded.

The way people, societies, cultures, and any other living systems are described reveals the labels and other abstractions that serve divisive agendas. How much context has been brought to the description? How much complexity?

We all know what it feels like to be misunderstood or categorized in a way that does not hold our whole selves. When people pick one context of someone's life as their identifying characteristic, the complexity of the person who is being described is violated. I am my DNA, but I am more than that, I am a mother, but I am more than that, I am my job, but I am more than that, I am my nationality, but I am more than that, I am my generation, but I am more than that, I am my body, but I am more than that. I am my religion, but I am more than that.

Even if someone's description of me could hold an accurate illustration of all of that, it would be incorrect, because I am not the same from one day to the next. I learn, I change.

To take all the contexts of someone's life into account in a single description is probably impossible, except perhaps through poetry or art, which allow for an unending shifting of perception. Describing another person or even describing oneself in a way that includes the complexity of life requires another sort of expression that comes from another set of assumptions.

What is communicated in decontextualized ways gets recontextualized into other stories. I may have been in the vicinity of a farm, but that does not make me a farmer. Or, I may have known someone who knew about a crime, but that does not mean I knew. Taking out of context one small slice of information and presenting it as a description of the person or community within other sets of stories is an easy way to recontextualize them. This sort of characterization distorts perceptions and impressions.

Schismogenesis

In the summer of 1943, Bateson, like many anthropologists, went to work in Washington, D.C. for the Office of Strategic Services (O.S.S.) as a "psycho-logical planner." His eagerness to fight the fascism of WWII led him there. The O.S.S. used his knowledge of culture and communication to confuse the enemy troops.

That office dispatched him to Southeast Asia, where he spent twenty months in Ceylon, India, Burma, and China. By his own accounts, he accomplished very little. The one activity which caught his interest, however, was operating a radio station aimed at undermining Japanese propaganda in Burma and Thailand. Studying its content, Bateson had translators engaging in small-scale symmetrical schismogenesis.

> We listened to the enemy's nonsense, and we professed to be a Japanese official station. Every day, we simply exaggerated

> what the enemy was telling people." Upon his return to New York, he gave Geoffrey Gorer the impression of being "very disturbed with the O.S.S. treatment of the natives. I think he felt that he was associated with a dishonest outfit. (Lipset, 1982, p. 174)

This technique of exaggerating and omitting information distorted the communication. For example, Gregory would intercept a message that might say, "The enemy is 37 kilometers away," and change it a little: "The enemy is six kilometers away." The immediate effect of tweaking this message is that the troops get ready hours early. But that is only the beginning. The disorientation of the communication has a knock-on effect of rotting the trust and relationships necessary between soldiers who fight and live together. This is what Gregory referred to as "schismogenesis."

This may be one of the most relevant and useful ideas for anyone interested in the runaway disintegration of socio-cultural communication.

The word schismogenesis means creation of breaking; as a communication pattern, it creates a generator function of communication that breaks communication. The confusion creates confusion, leads people to the next wrong question, and froths up distractions. Where an ecological system generates relationships that make more relationship, schismogenesis escalates relationships that break relationships. Today, divisive communication is eroding relationships and potential relationships everywhere. Even though this may seem like a trivial trick to most people, for Gregory, it was a deeply cynical act.

For Gregory, deliberately messing up communication between people is and was a violence that reaches beyond the bounds of war. It is a violence to the sacred act of communication itself. Of course, Gregory justified it during the war as something that needed to be done for all the right reasons (to fight fascism). There is a family rumor (this is not recorded anywhere that I know of) that shortly after the war, Gregory's despair after this experience was so deep that he tried to take his own life. His attempt failed. And he spent the rest of his life tending to and in respect of communication.

Communication is not what is said or even what is not said. Communication is what is received in the context. Additionally, communication is not what is on the transcript; it is what it was possible to say or not say. As living creatures, our perception is made up of combining bits of stimuli. We put notes together to form songs, we combine light to form colors, we combine context and words to form verbal and nonverbal communication, and we combine gesture, culture ... It's how perception happens. Communication is much more than words. It's gesture, symbol, silence and so much more.

Manipulating the combining of different pieces of communication to fragment it is a violation of life. Gregory ached with that knowing. Decades later, this is the core of my work on Warm Data and the rigor necessary to perceive information as alive.

How Does Change Happen?

But try ... just try ... to explain to the media or write a scintillating book about the importance of interdependent systemic life! Much of today's media will reject the warmth and aliveness of contextual description; there is not enough bite, not enough sting, and too many sides. The media likes a two-sided controversy, a basic polarity. The stories that hold our families and ecosystems in danger now have thousands of sides. The grave dangers of our time will not fit in binaries. But since the audience of the press is captivated by controversy, decontextualizing is justified; it sells. And in doing so, the ability to perceive the most basic form of life — relational interdependence — is suddenly an abstract intellectualization.

This is how you break the relationships that integrate a living system. Once broken, the parts of the system appear to be separable, controllable. However, the attempt to control eventually has the inverse effect, along with innumerable other consequences.

Given this understanding that "control" is at best an illusion and at worst a devastating violence to living systems, the question that Bateson was asking is: How does change happen? How does a family change? How does a marriage change? How does an ecosystem change? How does a culture change?

An understanding of living systems reveals that change cannot be produced through linear control — so how does one participate in the processes that produce systemic change?

This is a critical moment in the history of humanity to be asking that question. There have been many models of change and theories of change, and yet, those things that one might wish would have changed by now have not. Wars, corruption, destruction, the makings of the polycrisis are still very much as they were, only worse.

Gregory Bateson tried long ago to shift the habituated way of thinking about change as something linear. But the mechanistic dream of life being fixable in the way that a machine is fixable has proved to be a difficult dream to wake up from. Finally, after almost 100 years, it may be possible to begin approaching this question of change differently. Gregory's work, in particular, is just now beginning to be recognized.

How do you think about change if not in linear strategies? You tend to the relationships.

All living and social systems are vitalized through their relational ecologies. When the relationships shift, the whole structure of the surrounding relationships also shifts.

How do you tend to the relationship? Relationships are made of communication, even at the cellular level of your body, all the way up to global politics. So, if you want to shift the conditions of the relationship, you tend to the communication.

This is not about making a new script or vocabulary. It is about contextual and transcontextual shifts. Ecologies move in inter-relational ways — Bateson said this 80 years ago, and his father, William Bateson, said it in 1888. Indigenous cultures have been saying this for thousands of years.

Gregory was frustrated at how difficult this idea was to bring into a world of mechanistic habits. He died in 1980, feeling that his work was not understood. He was largely correct; cultural and economic assumptions are still stuck in thinking that is at odds with life. There is still little comprehension of how communication and relationship ripple into consequences untold.

Twisted Communication

Why? You may notice that this messing around with communication is a basic characteristic of daily life. Every advertisement is doing it. Algorithms, entertainment programs, politicians, news outlets — everyone is manipulating what can be perceived. The practice of twisting up communication is so prevalent it may even seem normal.

So why does it matter? It matters because interrupting, tweaking, lying, exaggerating, silencing, omitting — these little acts become huge fissures in our relationships. This practice dehumanizes, devitalizes, and shreds our deep connections to each other. The delicacy of life on this planet is hanging in the balance.

It happens at home, in our communities, and in the world around us. The infection of this breaking of our together-unity is tearing the fibers of life apart in the way cancer cells tear apart the body's ability to communicate. It makes the world a place where the only reasonable thing to do is "watch out for number one."

While it may be a prevailing logic of debating or convincing others of our position, communication distortion is a dangerous game. It leads to a moment much like the present, when it is impossible to know who is sincere and what information they base their sincerity upon. Who should you believe? What is left of our communing that is not for sale? A sentence is likely to be worded to evoke particular impressions. A montage of images can conjure a particular notion.

What would happen if there were a common recognition of communication as sacred, if it were not used as an instrument of control?

Of course, there will be mistakes and misunderstandings in communication. Some mistakes are extraordinarily painful, some are funny, some are confusing. This is different than strategically manipulating. But I wonder how easy it is to tell the difference in a world that is so accustomed to communication distortion as a methodology of everyday navigation through relationships in the workplace, school, home, and technology. I wonder if it is possible to find a way to allow communication to untangle from the pain and the profit of the last 3000 years. "Everything you say can and will be used against you."

Using communication to achieve a goal is likely going to backfire in unexpected ways. Communication is always entangled in many more processes of being alive than just speaking. The agenda may be clear, but the way it ripples out is not predictable.

Combining Contexts

I released a film about my father in 2011, after years of research and editing. That project pushed me to examine again and again my own limitations in creating a description of someone else's life, in this case, my father's. I knew I could not put every second of available footage in the film, nor could I account for the contexts around each of the decisions he made in his life. I am not certain that he could even do that if he were alive to try. I realized that in making a description of his life, I was describing my relationship to his life.

Again, this is a combining of my experiences and my perceptions with my own forms of communication. Is it fair to ask if the film I made about him is really about him at all? I could have included gossipy assumptions about who he slept with or highlighted his most epic successes and failures. It would have been a more traditional biopic if I had. But I found that tweaking of the information — exaggerating and omitting — made the story into something that did not give the audience practice in perception of complexity.

I found that shifting the blur of uncertain events into crystalized assessments of another agenda was like cutting him to pieces and reconfiguring him into someone I did not even recognize. The same schismogenesis that broke Gregory's heart is created in exactly this way. And yet, by not employing those techniques, the film was difficult to get into broad distribution. What this tells me is that the fast food of polarized gossipy reductionism is seductive and lucrative.

This misuse of one of the most precious aspects of being alive is ubiquitous. Our children are growing into a severely fragmented world, our bodies are showing the illnesses of disconnection, and our communities are divided. Politics and technology are merging into an alarming communication horror story with AI to categorize us all into our algorithms and pit us against one another on social media.

Communication is the birthright of living organisms to make combinations, contrasts, and generate relationships. To climb a mountain is to combine the body's balance, muscles, temperatures, and breath with the circumstances of the hike. Is it a date? Or are you running from a bear? We combine information beyond our verbal words. We combine impressions, we combine memories ... this is how living things make sense of the world.

It has always been important to be careful with communication, not to use it to twist or spin or augment ... but lawyers, marketers, advertising agencies, politicians, academics, pharmaceutical research, religious leaders, media professionals, and military strategists have been perfecting the art of manipulation of perception for so long, some say it's "human nature."

The Thirty-Six Stratagems, often associated with Sun Tzu and Zhuge Liang, lay out methods for psychological warfare thousands of years before anything like the CIA was ever imagined. Cambridge Analytica's manipulations of algorithms, and the mishap of news stories that cash in on controversy, reveal a collective selling-out of the sacred promise of communing.

There are some things that should not be for sale and prowess to manipulate communication is one of them. But alas, this is the world of business and exploitation. Beyond global politics and business, beyond our personal insults, beyond the place where "you are wrong, and I am right."

The revolution now is to honor a more alive world of relationships, and in doing so to honor life. Communication matters, both verbal and non-verbal. Sit by a fire, side by side, sing, hold babies, walk slowly as you assist the elders, grow, cook, and eat beautiful food together.

Time together in analog human-to-human communication cannot be underestimated now. There will be no community without first communing.

Alfred Korzybski (1933) said, "Thus we are living in a delusional world, a world of phantom semantic structures." The structures of language go on to shape our lives — not only in words, but in the world the words describe, and in the way we sit in silence together.

Fig. 41. Rachel Hentsch. (2026). [watercolor]

walls

Walls. Falls.
Broken things that cannot be put back together.
Kings ...
Schismogenesis (runaway system breaking) is in full bloom.

Machiavellian logic is eating its own limbs in a metallic tech-bro orgy of eugenic illusions of invulnerability.

Do not be distracted. The machine will use this chaos to pit us against each other.

Spoiler: Nature wins.
Totalitarianism is always temporary.
Destructive and terrible, but temporary.
Life cannot be controlled.

Does this breaking — like the moment an addict hits bottom — lead to more possible care and respect for each other and the beautiful blue orb of life?

It matters. The way. The tone.

The nature of the breaking matters to how it can heal and become a new form.

I understand that people wanted radical change.

Now it's here.

Erosion rode in on the blaming of each other — people, immigrants, women, gender, vaccine wars, religious differences, wealth gaps, health gaps — while justifying the ravaging appetites of big tech, big banks, big ag, big pharma, big education, big biz — thus double-binding the possibility of caring for each other.

Do you have time for your kids? Your partner? Your health? Your elders?

Time to break from any practical, logical reasoning that might suggest generosity is useless — a gesture toward making the day a little sweeter for someone.

Much change has been needed for much time ... and the systems of last month's status quo were not able to change from within. But a clamoring craving for forced change is unlike a move toward nourishing interdependency among living beings. The spirit of this frenzy is greedy, usurping, and dominating.

Each moment is a moment to stand in the full blast of intensity of our sacred membership in life. Lawless is the control and the loss of control.

 continued ...

Here we are.

I suppose the industrial road was always leading to this moment.

But this is not the end of the story.

Now the mess. It's rapid, but it's been forming for a long time. The propaganda machine has been very effective in turning people against one another — divide and conquer.

The revolution is to turn toward one another. Human-to-human.

Beyond the abstractions of voting parties. And to admit that no one was really here for the people anyway. That's up to us.

Five hundred years of divisions and pain are not easily overcome into a happy picnic. The alternative is unthinkable, and already in play.

Idealistic perhaps.
But worth saying.

We are in a storm.
It matters how we meet it.

Fig. 42. Vivien Leung. (2026). [watercolor]

Fig. 43. Vivien Leung. (2026). [watercolor]

the backwards wish

The black of the tree branches in autumn, moist and set against oranges, reds, and milky skies, reminds me of my mother's flower arrangements. They were somehow so elegant, allowing for the expression of each stem to shape a vibe. I never figured out how she did it. When she was 95, I would bring her flowers to arrange so she had something to do in the kitchen when she kept me company while I cooked.

The time came when the kitchen was no longer great for her (she would forget things on the stove, and her chopping skills slowed down). Her flower arrangements were always exquisite. She had an eye. Looking back, I can admit I was impatient and maybe a bit frantic; cooking for the whole family, working full days and often nights, and caregiving. The pace of her flower arrangements was not accessible to me then. I wish I had slowed down a little while we still had moments together.

And there it is, the backwards wish. The reflective wish that says, I missed something, that I now notice was important. This is not remorse; this is a most tender time-traveling gratitude for the opportunity to have learned something. The backwards wish was what saved my mother and me from eternal regret. Instead, I am here today, thinking about getting her vases out to honor her, and attempting to channel her uncanny grace.

We did not have an easy mother-daughter relationship. We fought, fiercely. We both considered giving up on one another countless times. But we did not. As she got older and closer to death, the possibility of her dying and our decades of resentment remaining jagged was painful for us both. We wanted to find peace and love between us to precede her passing. But we did not want it badly enough to fake it. Or maybe we wanted it too badly to fake it. We had the same fight over and over for 50 years. Eventually, we got through. Not everyone does. I am thankful.

She is gone now, but we are closer than we have ever been.

She liked to have a scotch in the evening. One night, about three months before she died, we were having the evening scotch, and our conversation devolved into our ever-looping fight. "You were never there for me." I would say. And she would reply, "Haven't you got anything good at all to say about your old mom?"

We went around and around, decade after decade. Some people are not afraid to be wrong, others evade the admission of wrong-doing at all costs. Despite piles of spiritual teachings and being a family therapist, my mother was not the sort of person who would ever say she was wrong. I wasn't waiting for those words. But that night it all opened, at last.

I heard myself say, "We cannot go back and change it now, maybe we could just say what we wish had happened. Knowing what we do now, what have we learned? What do you wish you had done differently?" I felt it as I said it. I was remaking our brew. We could change the alchemy in this moment. I could feel the eager cells of my body longing to hear her words. I was curious. An opening had arrived at last.

She handed the question back to me before making her own response. Typical. But, ok. I dived in. I told how I wished we could have had a home where we could live after my father died. I wanted to go to school and have dinner with her at night, I wanted help with my homework, I wanted us to live in the rhythm of the life I was in with high school. I wanted something like normalcy.

She was aghast. "You wanted that?" She then explained that she had been doing all she possibly could to introduce me to interesting people who had broken free of the daily grind. She was trying to set me free by dragging me to every ashram and philosophical retreat center in the world.

"Oh." I saw it suddenly. I saw her, I saw me.

The backwards wish revealed the context we could not speak to for 50 years. Tongue-tied in assumptions, we had never been able to get this exchange to happen.

From that night, we were on a new path. Six weeks later, she had a stroke, and three weeks after that, she died.

Our friendship is continuing.

Fig. 44. Nora Bateson. (2026). [watercolor]

grief

These are our times.

Give nothing less than everything —
    Hold-back is of the past.

Grief is contact
    with the holy ...

Masked in sadness
    Because the water of tears is coded in loss —

But the sacred is always beyond anything loseable.
    Water is fluidity of movement.

Loss is shared — and cries for the release from isolation.

eating the rot

After spending some time at a vulture refuge territory in the mountains of Spain, I returned to the rancid, corrupt political, economic world of rotten institutional systems — with new admiration for the giant birds who can eat the decomposing carrion that is poison to other animals. I have not known how to avoid taking in the anger, the frustration, the confusion, disappointment, and despair that is everywhere now (even in the eyes of the youth). Putrid centuries of epistemic justifications have formed into systems that set us all against each other, consuming the aliveness of the world. Extractive, transactional, death.

High above, the wide-winged birds circle, riding the air currents, coming in to clean up; only they can find nutrition in the necrobiosis. Below, cultural decomposition is harder to handle. I wonder, is this how a species extracts itself from evolutionary entanglement in symbiotic life?

Is all of this isolation and polarization, the breakdown of schools, health systems, politics, technology, financial systems ... is this how a species exits, by getting so entrenched in destructive abstractions that it's impossible to transcontextually get out of the traps? Is the human project rotting?

Might our species, too, learn to eat that which needs to be transformed?

Vultures have only a few feathers on their necks, and when they dunk their heads into carrion, they do not get soiled in the mess. They are formed in such perfection for their task. The disgusting and ugly stink that they ingest is the sacred work.

With an appetite for the noxious, they are noble.

The vulture eats the consequences of loss of vitality and turns fetid remains into nourishment. This is a repeated story of the making of a shaman, a witch, or a healer; to in-gest poisons, transform them internally, and out-gest nectars and elixirs of life.

Vultures do not kill for food. They have majestic talons and a strong beak that could easily serve a predator. Even bunnies know they are safe to hop around the giant birds without danger. These holy scavengers only eat what is already dead; their dignity is that of the necessary clean-up. Beauty and grace have many forms.

I would like to keep a metaphorical vulture nearby as I learn to receive the muck of this disintegrating time. I am learning from them.

Cooking a meal, tending a room full of people who I would like to experience whatever small grace I can offer in a day — this is the practice of every day metabolizing the rot of the modern world into a rebellion of noticing fields of relational consequences. Recognizing in every moment scripted ways of thinking that stink of exploitation ... these ingestions are sickening without the transform into family reconnection, a ceremony, late-night omelets, and whisky.

The vulture is a mascot for this time. Through the vulnerability of my own health, I have learned that I cannot directly eat the rot and transform it into life; I have tried.

The only way I have found to digest the cultural slime is by forming it into inquiry, poetry, story, irrational support of others, and erratic rhythms of giving everything and then nesting in the unflinching integrity of ancestors who refused to give in or give up on the infinite possibility of life.

Bring the vultures.

Fig. 45. Rachel Hentsch. (2026). [mixed media]

chewing ...

appetites and desires

The lion that hunts on the savannah has an appetite that exists within the relational dance of the organisms of the savannah. Though the lion's fierce slaughter of the antelope may appear violent, it has within it transcontextual ready-ing, or nth-order agency, or life-life-ing. Yet somehow, the appetites of human societies have become derailed from the relational vitality and healing of larger ecologies.

At the grocery store, a customer reaches for the most perfect, biggest apple on the shelf. That is most likely the least ecologically related apple of the bunch. The little apple that has not been engineered to use more than its share of water or nutrients from the tree is the one that will sustain the ecology in a drought or time of pests.

The wanting habits that have sought flawless, prescribed beauty have ushered in multiple industries — diets, makeup, surgeries, agricultural modifications, educational testing programs, psychological tests — all to further the measurement of "perfection."

The tension between wanting and controlling that which makes something wantable has soaked into the depths of multicausal processes.

Wanting more, wanting to own, wanting to claim. The hungry ghost is insatiable. No amount of wealth, or stuff, or beauty, or faster technology, or more status will ever be enough. And the attempts to feed these appetites create fragmentation in the transcontextual ready-ing process.

Yet the lion has no more control over its appetite than you or I do. In fact, controlling appetite seems only to distort it further. The cravings for, and ambition to acquire, these things are entangled in another complexity — not the complexity of mutual vitality, but rather the complexity of the interlocked social systems and institutions. These misguided hungers reach from professional ambition to sexual and intimate longing.

Consider the sensation of desire: impossible combinations of chemistry and culture meet a well-timed interaction to spark attraction. "I want you."

This could be mistakenly identified as first-order — a tragic confusion that begets the violence of exploitation and abuse. At this primal magnetism, there is more complexity than all the poetry and music in the world can ever express. Coupled with a cultural currency of having-ness, "I want you" often becomes predatory acquisition.

If we are looking for quick-fix answers and binary memes, we will find them, and they will not suffice to build new ways of life. But, if we can begin to recognize the complexity in our own identities, we may be able to recognize that of others and thereby humbly enter another level of mutual respect.

> Physical attraction and the body's pull to be close to another person is a simple phenomenon of life. Simple. That physical attraction is an expected experience that most adult humans have known. But, it is also complex. This mutual yearning emerges through combined complex systems, from culture to cognition and from the microbiome to material wealth. While attraction includes multiple, inexplicable ways of receiving and responding to the presence of another person, respectfully acting on that attraction requires consent. What is mutual yearning? (Bateson, 2023, p. 333)

In the most intimate experiences — of love, of loss, of sickness, injury, hunger, music — there is a constant seeking of the universal. Perhaps being alive, and therefore being within life, is in part a shared quest for the reflection of life's ever-becoming. I am alive, and my aliveness is confirmed in the living world. The longing for connection is a longing to touch the tension between being an individual and being within the contextual swirl of the cosmos simultaneously. Spiraling into our mutual ready-ing, all at once, very slowly, in impossible flickers of contact.

Like the many pollinators in a meadow, myriad forms of communication flutter, skitter, and ooze into other communication and spark into necessary interdependency. The tiniest and the closest of inclinations iterate outward, reflect everywhere, becoming a meadow of butterfly effects ... Every desire is resonating through time, culture, context, and communication.

An nth-order response is inevitable. In fact, nothing else is possible. Responding to the devastation and destruction of life as it unfurls in communities around the world will always be the result of nth-order responses. Some actions will confirm the effectiveness of the logic of destroying; others will open premises from which future actions reflect learning from the past, not to destroy life. Knowing the difference is probably beyond calculation — and perhaps that is the point.

To insist on certainty about which actions will do what is to flatten the very complexity that life depends on. But knowing that there is a difference — even though it can't be mapped or measured — is an attention to bring to each moment of the day, each interaction, each doing and non-doing. This is not the attention of analysis or judgment, but the kind of alert attention that notices the subtle reverberations of what is unfolding and allows them to inform the next movement. It is the recognition that each gesture — whether spoken, enacted, or withheld — does not stand alone.

Remembering the nth-order, ever-reaching, iterative, transcontextual movements of each action will intrinsically shift the landscape of potential actions. The field itself changes when we carry this awareness, because we are living in a way that is responsive to the living field. That is not a strategy — it is a mode, a way of being — a continual participation in the ongoingness of life, without finality. Our engagement is not to control but to keep threading in a way that tends to the whole.

Standing on the precipice of danger in this era is a confrontation with both the danger of doing and the danger of not doing. The vertigo is not welcome. The urgency to respond to the injustice and destruction is unquestionable. The urge to quench the vertigo is as compelling as the need to avert danger. Sitting around contemplating complexity can easily appear to be a capricious, intellectual bypass.

While complexity may feel uncertain and full of potential disaster, curtailing it and creating an idea of safety is far more dangerous. The temptation to foster safety from the lack of control is culturally supported at all costs, even at the cost of false safety, which may be more dangerous than any of the dangers we are collectively facing now. A shallow safety is not safe. The smallest of gestures also speak — sometimes scream — into the paradoxes of agency. Intimate and minute, the little communications carry the logic of the larger ones.

The runaway responses of life creating life creating life creating life creating life are side-by-side with the runaway responses of destruction making destruction making destruction.

Again, this is what my father referred to as schismogenesis: a creating of breaking. While relationships in ecological systems can create life, those same relationships, if pitted against each other, can also become destructive, polarizing, and break the infinite recursions that could nourish one another. This can become a runaway of relationships breaking relationships breaking relationships breaking relationships.

When global markets and petroleum prices lead to wars, and droughts intensify internal political strife, people attempt to migrate to other lands where they can feed their families. The wars then lead to widespread PTSD, and the trauma generates a need for pharmaceuticals and behavioral modification of veterans. Their children suffer further trauma from living with PTSD and act out in bullying ways at school. Parents, distraught because they cannot get their children to attend school — or pull them away from screens — experience stress at home that leads to trouble at work ... and the iterations keep going.

It may even be tempting to think of an action as a singular cause to which another action becomes a response. Yet perhaps it is more revealing to think of all actions as responses — and responses to responses — to responses, to responses — ad infinitum.

The compost offers a glimpse into the polyrhythms of transformation: food, nourishment, energy, decomposition, soil, and the conditions for seeds that join the overlapping temporalities of ongoing life. The generosity of life toward more life is present in the movement of every aspect of every organism's shaping, coloring, reproduction, consuming, nesting, rotting, becoming, seeding, growing, blooming, fruiting, wilting, and dying. Yet, this compost metaphor is more than a call for patience. The crux of the issue is that the transforming of the compost does not follow expected correlative lines.

The curious title of the earlier chapter, *Not All Onion Skins Grow Up to Be Conversations,* suggests that any necessary practices of attending to systemic movement will confront current notions of strategy as seemingly first-order causalities. While all actions are inevitably nth-order, all nth-order actions are also first-order at the point at which they are noticed. The difference is the basis under the action, the logic of the attitude and the approach, which, if there is to be an existential hope for living in sync with life's movements, will be a basis of knowing that the endpoint of an action is not within the provenance of strategy, but that all strategy is going to enter transcontextual interdependency.

There is no beginning or end to actions and non-actions. It may seem that a single action is a trigger for other ripples; a declaration of war, a surgery, the planting of a garden, a bee pollinating a flower — but those actions are grown through multitudes of backstories and follow into multitudes of futures.

What gestures, in this moment, are contributing to the composting, fermenting culture of ready-ing conditions for reaching into the infinite without violence?

A gesture of urgent humility is needed.

A gesture of grace remains everywhere, asking only for the most radical act of unclenching the fist of modernity.

A gesture of expanding the spectrum of expectations opens to reach far beyond where they have been captured in cages of direct correlation, and into sacred, nontrivial vitality.

Unspeakably, I witness the meadow, the forest, and the spider in the corner communing with all that I am — and where is the edge of me?

A mountain could be the speech of the god, or a river or the empire or the configuration of the stars. But, in the process of the centuries the mountain is levelled and the river will change its course, empires experience mutation and havoc, and the configuration of the stars varies. There is change in the firmament. The mountain and the star are individuals, and individuals perish. I sought something more tenacious, more invulnerable. I thought of the generations of cereals, of grasses, of birds, of men. Perhaps the magic would be written on my face, perhaps I myself was the end of my search (Borges 2007, p. 170).

Fig. 46. Noah Qwarfordt. (2026). [watercolor]

among-ing

Life cannot be rushed.

Several years ago, I had a vivid dream that began with a dusky, blurry vision of the moon flickering on water in the distance.

As I got closer, I realized it was not water, and it was not the moon.

What I thought was the flicker of light was a writhing of movement. It was a pile of robots with human form and flesh. They were sick. They were crawling and squirming in pain. I was so worried for them.

It became clear in the dream that they were sick because their designers had rushed to make them more human in time for production deadlines. These robots were just more products, delivered to market at scale and speed, prioritized above the delicate interdependencies of life.

Early programmers ignored the hundreds of thousands of years that it takes for a microbiome to commune with the ecology to make an immune system. The sick robots were infected because they were not able to be within the bacterial world. Those many generations of learning *among* organisms could not be hacked or expedited. Creating the form of a human body is not the same as that body's sharing of story within the many organisms of an ecosystem. "The map," as Korzybski said, "is not the territory." This error in coding became intrinsic to robots building more robots.

Interestingly, this is the exact opposite dream of the folks who have been hoping that by becoming more cyborg, human beings will become less susceptible to disease through enhancements by AI superintelligence.

What is this perverse appetite for the plastic safety of immortality over the earthy mess of living and dying in an eternal among-ness with life itself? Immortality, to me, is a poor substitute for being conjoined with eternity.

The dream viscerally landed for me the ancient credo shared by many cultures, that life can never be wholly understood and thus should never be underestimated. This is, however, not a bypass or a drop-off in the sense that one might celebrate the "unknown" and abandon inquiry. But rather an intensifying of rigor that invites complex, unending transcontextual movement.

Learning among organisms is not just data and bits; it is multicontextual, it swirls across time and species. A look at one's own body with this in mind may reveal a brilliance of elegant structures and processes that have been adapting to live in diverse (and ever-changing) contexts: in snow, in jungles, in inner cities, on boats, in deserts. And the learning is never complete but always continuing. Since everything is changing, it is possible to change, and possibility changes.

The robots in my dream were outside of the interspecies responsiveness that becomes immunity. They were left out, isolated, unentangled. Immune systems are conversations between symbionts and symbionts-to-be across and through a saturation of inter-steeping stories of being, surviving, and mutual learning.

Bodies learn to be in communication with bacteria and viruses over time, such that they can potentially coexist. In my dream, the designers of the fleshy robots were consumed by their own hubris and failed to consider the many organisms of learning that would be necessary to hold these new beings. The designers, distracted by the glamor of technological solutioning, thought they could skip the invisible depth of inter-being.

I awoke from this dream heavy with a deep sorrow and an unbearable weight of guilt for the sickness in the robots, noticing this is the same damage pattern that has destroyed soil and water, and laced breastmilk with forever chemicals. Reckless innovation strikes again and again. I felt responsibility for these nearly alive beings.

Who am I if I am not heartbroken by this scene? Does my own humanity rest upon the ability to feel remorse for these despairing creations? These are moments when the mirror is harsh; what does it mean to be alive?

Stories have been told warning of underestimating what it takes to create a life, and underestimating the way a living being makes attachment to its image, alive or not. The Greek myth of *Narcissus* is one such story, as are Mary Shelley's *Frankenstein* and Philip K. Dick's *Do Androids Dream of Electric Sheep?* (later adapted into the film *Blade Runner*). The replica is not a replacement for the way a being stores stories and experiences in their physiology.

Beware! The organism with embodied stories and experiences will make the mistake of projecting its own epistemic luster upon the blank replica and thus fall in love with itself. The modeling and porn industries learned long ago that the blank gaze is compelling because it allows the observer to indulge in their own illusions without the mess of a real person's complexity. Reductionism is attractive, and even addictive, precisely because it is missing the sticky, difficult information. This may be sexy on the screen, but it's fatal to the continuation of life.

Too many centuries of defining humanity as distinct from the rest of the living world have brought too much pain and loss. Witnessing these writhing beings in such suffering was to admit a terrible truth. There had been an oversight, a carelessness, an arrogance that now shines a spotlight on the need to be so, so, so, sorry. The only way through was to be crushed with anguish at having been feckless innovators. The spreadsheet and engineering plans did not include necessary shared experience of many organisms over time.

The human body is not distinct; it is among.

Mingling with the evolutionary shapings and re-shapings of countless other organisms, the human body is teeming with what I have previously called life-life-ing … among-ing. Perhaps this among-ing is more than interconnectedness, more than interdependence.

My belly, my body, my brain, my breasts are ever-changing amidst infinite stories of other organisms' struggles and learnings. The formation of nervous systems, digestive systems, reproductive systems, cognitive systems, is more than connected to the soil, animals, fungi, and bacteria.… They cannot be disentangled.

This is nth-order; this is the abductive process; patterning that emerges only when many layers of relationship co-shape each other. It's never simply A causing B. These organisms aggregate their histories.

Perhaps this word, among, opens a vista into the way of merging and blending from the cellular to global scales. Perceiving in terms of nth-order requires recognition of blending beyond nodes.

Among.

The word carries with it a history that brings together notions of being in the midst, mingling, with the temporal notion of "during."

Among also carries the physical notion of "kneading together," so that something is of its company or crowd. There is a fashioning to fit within other elements.

Among is a word that carries the possibility of decentralizing the individual organism and the chronology of time in its context. This is an invitation to what we have been calling nth-order, but with perhaps more inclusive movement and grace.

To be among friends displaces the position of the individual and foregrounds the group.

To be among the living, among the dead, among the flowers, among the ideas, is to be of them.

The kneading and mixing of the entities allows for there to be clumping, where finding the outlines of each becomes a different sort of task.

Noticing the outlines of a distinct organism among others reminds me of what is called a tilt-shift in filmmaking, where the focus sensor in the lens shifts to pull a selection of the frame into focus while it blurs the rest. This is a significant divergence from centralizing a particular point of view around which events and other entities are external.

Othering the rest of life is no way to live, nor is this a position from which to perceive the importance of being among.

How does one see oneself as an individuated organism while also being among? The practice of this stretch and blur of outlines for me opens the field where I might otherwise cling to nodes and nouns. What happens to time when the positioning of an observer is distributed among multitudes?

Interestingly, in one of the etymological descriptions of the word among there is a conceptual overlap with the word during. Again, the perception of being among is situated in a blurred dispersion, not only of position and outline, but also of time. Among the evolutionary changes over the course of the last two millennia, the chronologies of organisms in their environments is difficult to distinguish. This is the beauty of the way that the word among can mean during. Changes of climate alter the way organisms eat, nest, breed, etc., but the way organisms eat, nest, and breed is also altering the climate.

Among-ing during … time and causation are enfolding.

The way time moves in the idea of an among-ing brings the infinitude of nth-order into experiential memory.

Ralph Abraham (2006), who was one of the mathematicians responsible for what is known as chaos theory, based on chaos mathematics, used to make a point of saying that the reason for chaos mathematics was not calculation, but the reveal of that which cannot be calculated. In complex systems, the mathematics of movement in one part at a time is not useful to understanding change. Every aspect of every living organism is included in the living structures of many other organisms.

The stories of our experiences of being alive together have formed and continue to form our bodies, our communication, and our perception.

All of the ingredients change the way all of the ingredients express.

"Runaway" is a concept in chaos theory that deals with the uncalculatable. I was surprised to learn from Abraham (2006) that this was inspired by my father's concept of schismogenesis, a recursive breaking of relationships that further break relationships that further break relationships. When multiple contexts of disintegration are simultaneously escalating each other, this is an example of what is called runaway in chaos math.

One of the easiest runaways to point to, but most difficult to grasp, is ecological runaway. The oceans' currents begin to change, the marine life is poisoned by plastics and heating water, the warm seas shift the winds that change bird migrations altering the movement of seeds; the soil is bereft of minerals, and fires, floods, and droughts are more frequent; economies are pinched by the loss of water, grain, and transport; political parties begin blaming the other parties; social divides take hold, crime increases, wars begin, cancer rates soar, poverty is rampant, mental health is in decline....

All of this is because of all of this.

There is no linear causal thread to either follow or untangle. This runaway has been called a polycrisis, a metacrisis, or an omnicrisis, but no matter what it is called, it is not controllable by any singular measure of containment. This is what chaos math can show us. Runaway reaches from the microorganisms to international politics, wrecking relationships on a global scale while severing families and dividing neighbors who might otherwise help one another.

Bringing the concept closer to our human experience, another example of runaway might be a scenario in which the doctors say there is nothing they can do after multiple organs in the body begin to set one another into crisis. When the digestive systems fail, and build-up of fluid in the body due to lack of circulation begins to prohibit breathing, the panic of loss of breath sets the heart into high pressure, the nervous system goes into overdrive, the microbiome is overrun by candida, the cognitive functions shut down.... There is runaway.

All the systems are disintegrating all the systems.

But this is not the only way to describe runaway. Runaway is also how life makes life at nth-order. Instead of the aggregate entities disintegrating the connections between each other, another possible way to look at runaway is to notice a meadow or an ocean as a runaway of vitalizing and communicating, interacting and entangling that brings momentum to integrating living organisms.

My story changes your story. Your story changes my story. These shifts can bring shifts that bring shifts to shifts moving either toward more vitality or shredding those relationships that make more relationships.

Runaway can make life and unmake it.

The aliveness is movement among, and this can go in many directions. In a forest, there are thousands of years of decomposing organisms that are making the soil in which the roots of trees and other plants can find nourishment, while they, in turn, provide homes and food for the organisms above ground, who also feed and fertilize from the same soil while they keep the various species from overshooting. The birds carry the seeds and poop them so species can reach into other places. The insects, bats, and pollinators are all making one another's lives possible in relation to the flowers that depend upon their pollination. Silica and minerals from the organisms feed the bacteria that feed the soil.

All of this is alive because of all of this.

There are more organisms in runaway in the continuing of a forest than I can begin to list, but trying to attempt such a list is a practice in awe, not spreadsheets. I am awash in gratitude that such fantastic complexity is right there under my feet and oxygenating my body as I take a walk in the forest behind my house.

Frequently, runaway is referred to in terms of acceleration of compounding situations. While acceleration implies a speeding up in linear time, it is perhaps better understood in the arcane sense of *quickening*, which was not merely speeding up but becoming alive. In this sense, the concept of runaway is a key patterning to study in life sciences, which are by definition among one another, and quickening.

The word *quicken* originally referred not to speed, but to coming alive. Its earlier meanings included "to give life," "to come to life," and even the first stirring of movement in the womb when a pregnant woman feels the child begin to move within her. Only later, during the industrial era, did the word acquire associations with haste, acceleration, and increasing speed (Online Etymology Dictionary, n.d.).

This shift in meaning mirrors a larger cultural shift: from understanding life as an unfolding process of vitality, toward understanding movement through the logic of production, efficiency, and output.

Quickening was once about aliveness.

This makes so much sense, looking at how the intensifying of life producing life, as a nonlinear becoming, could through the lens of a factory become reduced to speed. This was the fate of the robots in my dream. They were made in haste, not in the quickening of life, animated but never among.

Quickening describes non-linear, living, and aliving motion in a way that brings it back to the belly, and out of the production line. Almost every ancient culture has a version of the motto that there is not enough time to be in a hurry. The makers of the robots in my dream further exemplify this universal truism. The robots become a counterfeit quickening in a movement without being among.

Acceleration is a tricky concept in a complex system operating at nth-order. The speed of change depends on relational possibility. Each organism moves in different rhythms, making their aggregate movement in nonlinear time. Acceleration as a measurement on a linear timeline obscures the multiplicity of temporal relationships in play. The pace of this movement is among the organisms in combining rhythms and patterning. This is the runaway of relationships that make relationships.

Yes, it is the mystery, yes, it is unknowable, yes, it is incalculable.

The life-producing version of runaway can be an equally overwhelming concept, as it essentially refers to the compounding, exponential momentum of multiple aspects of coherence in a system like the body, or a family, or a forest, integrating into one another. I have referred to this as *simultaneously implicating* in other texts. The deer's antlers are implicated in thousands of other organisms, as are his hooves that mulch the turf, his poop that fertilizes, his microbiome, his fur, and his teeth. Where is the edge of the deer? This is the coalescing of unseen processes that make vitality — a becoming through combining that I have come to call *aphanipoiesis*.

Returning to that moment in pregnancy when the mother first felt the movement of life in her belly, I remember the awe I experienced when my babies moved around inside me, reminding me that I am not making a life, but that I am inside life as it wondrously continues. In that moment of feeling the quickening inside me, I was with the bears, the deer, the lioness, and every other mammal that has ever given birth. Inclusion in this infinite folding of life is nothing less than sacred. But exclusion is devastating.

The opposite of among is isolation.

There is no aspect of life that can maintain in isolation — not a blade of grass, or a bacterium, or a horse, or a child, or country, or a tree. Nothing can live without among-ing. Isolation is fatal.

Many creatures appear to share what we'd call the inclination to suicide, including dolphins, whales, other primates, dogs, and cats. The conditions that usually lead to suicide in these creatures are isolation and/or captivity. Being trapped and alone completely and totally disrupts the innate need to be among. The robots in my dream were isolated from other organisms at the cellular level but placed among them in their bodied forms. They died of exile. From this fear of isolation, the need to be among can distort into weird habits and defenses. One of these is the misconstrued need for (false) safety.

The most terrifying thing a parent can say to a child is, "I don't know," and simultaneously, this is the most reassuring statement a parent can ever make. A parent who can say they do not know what is going to happen is a parent a child can count on to be honest with them when there is danger.

The communication in intergenerational relationship is the basis of mutual learning, the basis of possible relationship with the parent, and the basis of being able, also, to not know. That child is not alone. To "not know" together is to be alive and learning together.

The reassuring parent who pretends to be in control offers a false safety that the child will soon discover, learning henceforth not to rely on the companionship of that parent in times of danger.

When injured or endangered animals such as condors and mustangs are restored to health before being set free in the wild, they must also be reminded of what it is to be among life again. Domestication is another set of rhythms that produces a false safety and undermines the animals' ability to improvise in context. One of the issues is that these animals become accustomed to feeding schedules, and when set free, they get anxious when they no longer have a predictable food delivery supply or system.

They also do not know what to fear. To live among other creatures and environmental conditions requires a healthy fear of the appropriate situations. Animals that have been in captivity become confused about what to do in storms or when they encounter a predator.

The wild is unpredictable, and this can leave a creature accustomed to being in captivity to become disoriented, isolated, and harmful to itself. Additionally, of course, the captive creature has an under-informed immune system.

> A wild population of any species consists always of individuals whose genetic constitution varies widely. In other words, potentiality and readiness for change are already built into the survival unit. The heterogeneity of the wild population is already one-half of that trial-and-error system, which is necessary for dealing with the environment.
>
> The artificially homogenized populations of man's domestic animals and plants are scarcely fit for survival.
> (Bateson, 1972, p. 458)

Runaway is chaos and confusion. It is incomprehensible, incalculable, and the variables are infinite. The among-ing of stories present in a war, or an addiction, or a bowl of soup, or a cup of tea offered by a loved one, is not possible to chart.

While this confusion feels like something to fear, control, and avoid, it can also bring mind-boggling astonishment. The same confusion that defies the journalists, the doctors, and the parents who are tasked with knowing what is going on … that same confusion holds a universe of mingling stories from every organism since ever.

Confusion in another form, another angle, can be inclusion in the awe of being alive amongst all of life. In this clouded fog of fuzzy outlines, there is bedazzlement, astounding vastness, and warmth to nourish the morale needed to anticipate the un-anticipatable events now gathering together.

Holding each other among and keeping a keen eye on those that are slipping into isolation; for their un-among-ing is my, and your, agony too. Beyond what I might have once called liminal or in-between, where the edges of the peaches and the bees and the summer harvest are difficult to parse, their life is among-ing.

Awe in the belly is the fire to love, and the humility to know life is breakable.

But what should a person do?
Will among-ing leave us in blurred nihilism?
Or is it enough to be alert to the among-ing and carry on from there?

Will the warm data that among-ing reveals shift the logic under the motivations for doing not just anything, but everything?

Warm Logic is a logic that welcomes the movement and variables of runaway life. The way one approaches everything shifts with this sensitizing to among-ing.

And yet the itch for a how-to persists. Yet, if I lay out any method of what to do and how to do it with proper Warm Logic, we are back at the recipe, and we've lost the artisanal practice of crafting each moment in context.

Can I feed you that?

Fig. 47. Nora Bateson. (2026). [watercolor]

second-order learning ...

There is an artistry of stitching contexts together.

Without mentioning a single theory, my father taught me to perceive complexity. Before I could speak, I learned to notice abductive process, and soon learned to commune in second-order cybernetics. These are not subjects that should only be learned about. These are not subjects at all. They are the rhythms and patternings of a way of being in life.

So, how did he teach me, a little girl with no academic framework to reference these theories, to inherently, intrinsically, and instinctively attune to the movement of systemic aliveness?

As I learned much later with my own children, the grace of my father's parenting was not so easily repeated. I have tried, and at times witnessed remarkable mutual insights, but more often I have felt my culture's script prevail. The grace can slip away, and teaching can become a humiliation of an individual instead of an enlivening of a group. There is much to learn about learning.

As a child, I was invited into an aggregate of stories and examples that spanned multiple contexts. These ranged from 19th-century vinegar advertisements and stories of dolphins learning to the way that leaves and stems have buds in their angles. The diversity of anecdotes and allegories was not immediately coherent to the outside viewer, but that was the stretch. I was invited to witness overlap, to inter-implicate, and to find my own experiences inside the arrangement of descriptions. My father rarely gave me a first-order, direct lesson, and when he did, he was usually being ironic.

Each day there are thousands of unspoken messages that imply how a person should learn to be in their family, school, workplace, place of worship, or town. I teach my kids to do the dishes and to be responsible with their money so that they will be prepared for life as individuated citizens in societies that value independence.

How might I relate differently to my kids if I knew they were never leaving home, as it was in other times, and still is in many cultures? What would the implicit messages be? I imagine they might center more around communication, and covering for one another, such as attuning to the relational information at the dinner table.

Getting in trouble for not doing homework is an artifact of an entire culture of intergenerational communication that fortifies the future of an isolated adult. How might that undermine teaching compassion or systems thinking?

The underlying principle behind my father's style of teaching reflected the way that life is always borrowing from previous communication while assembling mutual learning into the next communications. He taught me about ecology by revealing ecological patterns. From the outside, it appeared nothing like any instruction on anything at all, let alone the systemic complexity of living ecologies.

Classrooms in the larger culture of schools are typically knitted into socialization in the family, so that teaching and parenting have a similar tone. The "teaching" moment has something to do with "telling someone something they need to know." However, in our household this was held very differently, akin to the way an organism evolves in relation to multiple other organisms. The direct demonstration was more like a flower arrangement, where the lesson is not in any of the individual flowers but in the way they are placed in relation to one another, which shows their colors and shapes in new ways.

Communication worked as notes in music that are expressing one another, or spices that are combining to form different flavors — ideas and stories as an archipelago of concepts. This style of teaching presents the organism, human or otherwise, to a second order of learning as alive as life. There is openness to stitch contexts and restitch them over time.

This is always true. There is no way to escape implicit second-order learning. The question is what is placed in direct communication and what realms are left to make invisible, implicit connections within. A student in a math class is perhaps learning a little math, but more than the math, the hidden curriculum is a potent and insidious set of cultural expectations about identity and success in a world that values math class. The student in the math class is also learning what it is to be seen as an individual, to be in competition with other individuals, to be measured, labeled, socially placed among their peers, and to be assessed by the institution.

What would it look like to reverse that notion of the hidden curriculum, such that the tacit transmission between teacher and student is one of learning to be an interdependent organism in an ever-shifting world of symbiosis and ecology?

What would it look like if the underlying assumptions were of being among, instead of being in isolation? What would it look like if the implication of the courses placed the students in a world of ecological thinking? This is much more than knowledge about ecology; it is a world of movement and inter-shaping that defies the habits of industrial epistemologies.

To teach in a classroom in this way is to be an artist. My father was an artist. An artist does not depict things as they are, but rather works through a digestion of impressions and experiences. The artist reveals a description through which others may reflect a multitude of perceptions, memories, and ruminations. The artist provides a scaffolding that others may fill with their own architectural imaginings. In this way, the artist is always learning along with their audience.

Mutual learning among teachers and students, artists and audiences, parents and children, is an experience of being living organisms in a living world. I have often cringed at the violence of the thought, "I am going to teach that person a lesson." Nothing good comes from that position — just further schismogenesis, humiliation, and isolation.

With a twinkle of mischievous delight, my father would often recite Wallace Stevens's poem "The Man with the Blue Guitar":

The man bent over his guitar,
A shearsman of sorts. The day was green.

They said, "You have a blue guitar,
You do not play things as they are."

The man replied, "Things as they are
Are changed upon the blue guitar."

And they said then, "But play, you must,
A tune beyond us, yet ourselves,

A tune upon the blue guitar
Of things exactly as they are.

(Stevens, 1952, Canto I)

I am not advocating for abandoning rigorous education in disciplinary knowledge. Rather, I am invoking a counterpoint to the traditional education practice — a way of learning from within an epistemology that echoes the way a forest, or an ocean, continues and changes in relation to other living systems. The bits of knowledge that are transferred in a scholastic institution are of some value, but the contextual implications of the learning are what shape culture, identity, and possibility. Neither can be overlooked, and the latter is vastly underestimated.

It is well known that the school system and the factory mirror each other in terms of their purpose to efficiently produce product for the socioeconomic system. Disorientation from this long-held directionality is necessary, and confusing. How shall there be a new orientation if there is not first a disorientation? Wayfinding in open waters is terrifying to those who have only navigated with electronics.

Control is the central undercurrent of isolating subjects, making knowledge static in time, and providing learning as "here is the material, learn it."

In Phillip Guddemi's book *Ecology of Mind — The Course* — a collection of transcribed lectures by my father from his course, An Ecology of Mind at the University of California, Santa Cruz — I was able to revisit the way my father taught. His placements of various vignettes and examples are orchestrated in tandem with one another, such that the students might get a glimpse of the way the contexts intertwine. Alas, many of the students missed the gist of this and were anxious about whether the seemingly disorganized bits and pieces would be on the test.

The students had been taught to be in the classroom in a particular way: to take notes on what the teacher said, not on what was connecting for them along the way. The point was not the point as they had come to seek it. The point was exactly that the point was elsewhere, and in constant shifting. The point of what my father was teaching was that grasping for the point will obscure the necessary vitality of perception, which eschews points.

To suddenly ask teachers to be artists and to treat classrooms as ecologies is not within the current purview. The looping structures that bind coming generations to the flaws in the previous generations' understanding of life are tricky to untangle. Likewise, telling students to collaborate or think ecologically — in a context where they are measured on individual performance and production of work — is exactly the sort of first-order control to which modern students are responding with frustration.

If I pose the question to you: How would you teach in such a way that the content of your teaching was not the material itself, but what it implies when contextualized? How would you deliver material at first-order that invites entry at second-order into a world that has always been one of symbiosis? And how would you guide other teachers?

This is how a composer, a cook, and a painter meet their audience through their media. Individual notes don't really matter, the ingredients are not the highlight, the dabs of blue or red paint are not the painting. These elements make room for the recipients to do the combining through their own living filters of perception.

Imagine if we were to study a composition of blues music by pulling out all of the A notes, the B notes, and the others, and counting them up? Imagine if you piled all of the different colors in a Monet in separate heaps? Imagine if all of the ingredients of the soup were just placed on the table?

To bring this sort of artistry to the classroom is, I believe, to invite students into a practice of sensing their own complexity and that of the world around them. No amount of vocabulary about complexity can do this. The learning must arise from within the complexity. Otherwise, the students will be drawn back into competition for a better grade on the bell curve in their class on complexity thinking. The irony is profound.

When I arrived in the field of teaching systems thinking, I found myself in a paradox. I noticed that if I assumed the traditional stance of instructor delivering course material, the room would leave with a little more vocabulary from the discipline of these theories, but probably very little new understanding regarding their own lives as complex transcontextual stanzas in the poetry of life. However, if I were to adopt the trendy idea of experiential learning and gave them outings and group activities, I was also somehow transmitting the same self-perpetuating forms of classroom etiquette. The exercises became examples of the vocabulary, instead of tacit, untraceable presuppositions.

However, when I provided, through story or art, some patterns of life's movement, it was seen as vague, and off topic. The form of serving a smorgasbord of ideas and stories for people to tissue together was a familiar form for me. This book, *Belly*, my previous book *Combining* (2023), and my first book, *Small Arcs of Larger Circles* (2016) are compilations to be enjoyed in any order. The seemingly disconnected pieces in each book yeast an understanding not mentioned in any of the texts, but arranged to ring and sing among them.

I have received criticism for this unorthodox approach, and that is okay too. The shift from teaching about to teaching from within is a radical communication and meta-communication shift. Likely, it will not be visible to those looking for explicit definitions and linear outcomes. This shift in teaching is dangerous to the deep underpinnings of modern society and industrial colonial habits. As such, it upends the crux of what education has been understood to be.

Good teachers have always been storytellers. They have always known that the matter to which they tend lies deeper than the surface, the facts, or the details of the subject.

Fig. 48. Vivien Leung. (2026). [watercolor]

flags

The wrong word can burn an evening, it can sour a friendship, and poison a conversation. Perhaps that is the attraction of speaking in well-rehearsed scripts, signaling common ground and mutually recognized as familiar. The script of political views, the script of parenting approaches, scripts of spiritual alignment, or scripts of individual accomplishment, all of these scripts wait, ready at the tip of the tongue for their turn in the dialogue. They are reassuring, and confining. They are loaded with signposts defining position, level of education, culture, and attitude. The words cue solidarity, and they trigger polarization. Each word spoken is a world of communication beckoning histories and futures. All words are spells, spoken and unspoken.

The intensity of choosing words is for me an alive and activating project. I feel them form in my mouth, finding sentences, ringing and pinging off the walls of a conversation. Each sentence is a mouthful of flavors. Spitting flowers, spurting cliches, slurping idioms; making words, writing words and saying them aloud can be an invitation and a violation simultaneously. Increasingly, the tickle of language in flux has brought me much delight, and also, sorrow. Delight in new possibilities for perception, and sorrow in the sharp blades these changes in language have brought to communing.

Beyond the reach of words there are galaxies of other communication. But, within language, words are what we have. They are hell, and they are our glue. Sticking us together, sticking us to our past, sticking us in grammars of exploitation, idioms of colonialism, metaphors of modernity.

Words are traps, words are portals, words are stones, words are liquid. The spells and incantations of day-to-day conversation reach into wells of complicit meanings, even in arguments. They form trends, fall into grooves, get stuck on repeat. Different eras favor different grammatical patterns.

Words change.

They are glue and they come unglued. Thankfully, language is alive, absorbing and metabolizing cultural assumptions. Communication cannot be frozen. When language changes, culture changes, shared presumptions move; perception morphs. The world is altered when words are altered. And they are always altering, this is good news.

In a time when so much feels so intractable, language is always adjusting as perception alters. There is no way to trap the vernacular into something static. As I write this I am editing, replacing, adjusting the language. I am asking, "Is this what I mean to say?" or "Is this the tone I want to say it in?" Have I entered the ecology of communication with my bare feet in the soil, or have I poured bleach on the new sprouting seeds? Am I seasoning the sauce or scorching it?

I worry, I care, I screw up, I worry, I care. It matters. Our communication in this moment is like a time octopus, reaching eight directions into our lives of before and after and overlapping them. So, I try. I learn. I shape the words in my head, I hold them up to the light, I see if they glow in the dark. And still, I know that this is not a formula, there is no way to get it right. The whole universe of language mocks us all each time we think we can capture it. Do we use language or does language, which is alive, use us?

There have always been recursive loopings around what can be perceived and what can be named. Is it only possible to name that which is perceivable, or is only possible to perceive that which has been named? If something is named is it then more likely to be addressed? Or does addressing an issue directly by naming it ossify it and push it out of reach of the magical underworld of unseen movements where it might be shifted without having to be made obvious?

I see the possibilities of learning about old words, making up new words, as an entry into communication and relationship that is shiftable. Unsticking the way we are together by ungluing the discourse that hosts our perception. I like to put odd things together, and pile uncomfortable contrasts between words and metaphors. The confusion is culpable, I admit that, but the stirring is necessary. The disorientation is an invitation out of the guardrails. It is not safe, but nothing is safe. I may be crucified, but it is the way to rebirth. The words will jackhammer the concrete sidewalks of expected conversations, and in their rubble tiny sprouts will poke through. I see the play and movement of language as an opening to the light, and the darkness of unspoken, unnamed potentials of relationship.

To learn, to move the sentence is to surprise the words themselves. To un-sentence them. It is like sticking my hands into the Milky Way and swishing around the galaxy of stars. So much is there.

But it can also be a sentence, incarcerating meaning. The alter of what cannot be altered.

What if every word, every gesture has become a flag?

More than buzzwords and memes, diction is a reveal of where effort has or has not been made toward excavating the baggage words carry. The community knows its mission statement, and the words are its mascots. They become sacrosanct artifacts of unification, and then they become weapons of division. How many versions of this story have we all seen throughout history? The rise and fall of empires, religious leaders, activists, heroes, intentional communities, love affairs. Some contracts are worded to constrict, or, maybe all contracts are constricting? The promise becomes the problem.

Each communication lands in a context rife with polarities in all directions ... starting long ago, and exacerbated by media algorithms, we are pitted against each other. Even in the name of collaboration, there is competition and isolation. Be careful with flags; they are dangerous banners of division. Some flags will unify some people and gain applause, but others will contradict solidarity and ignite vitriol. Everyone becomes a bully or a fan. Fans can be more dangerous than bullies ... and communication of almost any sort now becomes a rupture to the possibility of deepening our ability to be together, exploring the complexity, finding new pathways to co-exist.

I love to learn about the history and nuance of words. Their etymology and the secrets they are carrying. More often than not, common language is smuggling nasty ideas into fresh moments. I know I cannot help but assist in the smuggler's crimes, but I still want to know what I am smuggling. Mostly I find configurations of control from other times.

Again, words like "facilitation" (to make easy) encourage hijacking the vast saturation of learning necessary to alter perception, skipping over the hard work, hiding ugly histories. Beware the insidiousness of accepting language casually, lest sloppy understanding should become an entire modality of being.

To criticize someone's flag can be a violence to their identity, and to the community in which they have belonging. The word, facilitation, is a flag, one of many that populate conversation in an average day. And yet, the words that have become flags are all too shallow to hold the complexity needed to meet this moment. There is nothing that can be spoken of without requisite depth now. In these times, each experience is ripe with too much systemic overlap to swallow mono-topics being fed to them on PowerPoint.

There is vomit in everyone's mouth. The flag, like the recipe and the hankering for the cheap ease of a formula, all refer to important aggregates of details earned in toil and worlds of relationships, without taking the time to rehash the difficulties. Much is lost. Vocabulary references experience but can never replace it.

There is new danger in the old habits of looking for sovereignty in the cues of rehearsed scripts, jargon, and lighthouse words that signal belonging in one group or another. As technology gobbles up language, it is digesting it into cardboard. The words that enter the belly of the AI beast are rendered empty almost immediately upon their arrival in digital form. Experience is stripped from the words. They begin to mock us and make our verbal descriptions into crappy clichés.

For so long, those scripts and jargons have been harbingers of affinity, kinship, and loyalty to particular causes; now those same scripts are the susceptibility to repeating empty digitized grammars made without the seasoning of physical touch. The scripts are vapid shells of communication now. Beyond being boring, these well-worn signalings are polarizing, othering, and bring the chaos of artifice into precious moments of communing.

A grunt, or an ooooh, or an exhale, or a new utterance the mouth finds, are beyond the vacuuming up of language by computer processing. Can you taste the weirdness in communication that has been instructed by the synthetic wisdom of AI? If you speak in AI suggestion, and I speak in AI suggestion, are we speaking?

Analog communication is human-to-human, human with creatures, human as creatures among creatures ... To tell stories, to go together into the forests of our confusing world, and find our granular, minute openings of communing. Not just minutes, but time, deep time, to wander, in and out of multiple contexts, multiple stories together.

To try new language together is to travel into landscapes unknown.

The script of before is like a map of another place here, a recipe for another feast.

life wins

Spoiler alert: life wins.

There is little evidence that life as I once knew it can abide this trajectory of encroaching surveillance and ubiquitous technology. Interference in the subtle systems of nth-order communication is untethering the human body from life.

So many infected systems are in runaway; it is unlikely that there will be an untangling into utopic resolve. But apocalypse means uncovering, revealing.

Even in the bleakest of dystopic visions there is unevenness. Even when the big asteroid swallowed the planet in darkness, our grandmother shrews made babies.

A few protozoa and their friends will make it through whatever destruction Homo sapiens wrought.

A few wiggly little things will form a blob into which all of life imbues.

Relax — most of the attempts to "save the world" have been deeply misinformed by ideas of dominating nature, and only wreak more havoc. And yet, I am nature, I have children, and I make ceremony. I cook, care for ,and pour my efforts into the possibilities of each day. Hubris and humility. The microbes are on their own journey.

I may not have grandchildren, or great-grandchildren in human form, but the microscopic organisms are still of me.

The protozoa may be the only living improvisers left.

In which case, everything I do in a day now is for the protozoa. I am on team protozoa; to be so is to honor completely that no form of control over life has ever lasted. There may be great damage to many species, but in the end, life wins.

The controls rust and dilapidate. The weeds will overtake the data centers, digesting them back into elemental vitality. All the hubris will topple. A few protozoa are all it takes to get the quickening going again. Soon enough, the crumble of old structures will be climbed by vines with sticky stems. Then other little lifeforms will find a way in. Then larger ones …

As billionaires become trillionaires with very opulent bunkers stocked for decades of surreal canned meals served in the last gasp of the illusions of crazed culture, it may become obvious that addiction to having everything was always a losing concept.

The loss is therefore for what?

Well, that is a question worth asking.

Because anyway, it will be up to the protozoa.

Fig. 49. Rachel Hentsch. (2026). [watercolor]

trail angels

Nourishing the unknown.

Trail angels are people who place food and other items on long walking trails, like the Appalachian Trail or the Pacific Crest Trail, for hikers to find. These angels are often anonymous, and they do not know who will find the things they leave. Sometimes they leave coolers of drinks or snacks; sometimes they leave medical supplies, clothing, boots, or other items they imagine the hikers may be happy to come across. For many trail angels, this gifting keeps them involved in the hikes, even when they cannot walk the trails themselves for whatever reason — work, health, or injury.

Of course, there are scenarios where this goes awry, and the trail angels' gifts are unwelcome, or they are distorted into a transaction somehow. The tradition is not free from the contaminations of capitalist culture. But, for the most part, the spirit of the trail angel is to leave something for someone who does not know there will be a gift for them, to make their hike more enjoyable.

Hikers talk about "trail magic" — finding socks just when theirs were frayed, or blister bandages when they needed them, someone turning up at just the right time and giving them a ride at the exact moment they ended up on a road. It happens in towns and cities too — a blanket left on a highway overpass, a sandwich where you have seen someone sleep. People, doing nice things for people, making magic.

The metaphor of trail angels can be applied when thinking of the difficult questions around what can be done now for future generations. Is there a way to leave something on the trail for them, to give them morale, to boost their strength, or just to bring some joy? To offer something, with no knowledge that it will be received, or who will receive it, is another sort of offering.

Trail angels cannot be sure what the preferences are of the particular hikers who come across their packages. Some hikers prefer sugary drinks for the energy, others shun sugar, or favor spring water, or other unsweetened beverages. The trail angel can get it wrong.

What can be put into place now, during the current systemic habits, that will be welcomed by those who are living in another sort of world?

A world that is not possible to imagine now, a world you and I may be long departed from ... what is it that can be placed on the metaphorical trail for the people who will be here later to find? Just under the skin of this question is a world of rebellion.

The idea that it could be a reasonable thing to do, to leave ideas and possibly structures in place that seem incoherent today, for a world that may or may not appreciate our efforts. This can be seen as a nonsensical waste of time and money thrown toward something that potentially goes nowhere and cannot be assessed as having been useful later. There is no way to know, and that is the underlying essence of the gesture.

Precisely in the impracticality is where the rebellion lives.

This is the contrarian response to all things calculated to be "worthwhile." The action is not charted on its return on investment. It is wild.

The character of the giving is in contrast to the transactional logic that has produced the consequences of what is sometimes called metacrisis, or multicrisis, or polycrisis. These ways in which so many contexts of participating in life have been entrapped in existing systems that isolate, reduce, and destroy necessary human entanglement with one another and the living world.

A gift placed into the world that I won't see is a signal that — however confused and misguided the gift may be — there was someone in the past who tried to leave a trace of goodness.

If I imagine a world wholly upside down from the day-to-day living of now, what would I want to add to that? This is an nth-order gift. An action of actions; a doing that begets and rebirths others doings, reaches the now, and the past, and cups into the hands of people I will never know.

Of course, this question is also one that recognizes that the timelines of systemic chaos and decay are not evenly distributed. Many are already living in a future in which all the existing systems have failed them. Few people are held up by the institutions of the last few centuries. The basis of the project was exploitative from the beginning. The trail angel is a nano-disruption to that basis.

If I am asking how I might nourish a world I have never met, I am faced with my illusions around nourishment. Reading labels in grocery stores is surreal. How many grams of protein, how many grams of sugar, how much fat, and what are the daily allowances? These labels reveal a world in which food has been taken out of the context of seasons, cultures, gatherings, fire, story, and old ways of living in place. Appetites have become attracted to hyper-sweets and confused by synthetic flavorings and colors. I am not sure what I want to eat; the intelligence of my cravings has become disoriented. Perhaps there is a way to leave some form of reorientation on the trail. What would that look like?

What might I put my efforts into now, that our grandchildren's grandchildren might need?

There is something refreshingly humane in the trail angel, something to undermine the feeling of impossibility of escaping the patterns of the modern world. A gentle upending of the tit for tat. Opening the gates, recognizing that I can't pick and choose what I want to see in the future.

The wildness is in ironic opposition to schismogenic divisiveness. But the trail angel is also not a savior, not a helper, nor a do-gooder with an altruistic measuring stick for god and tax returns. The trail angel is living into the adventures with the hikers, not for them; participating in the unknown.

I have been in countless board meetings where the one person who was willing to negate all concerns for externalities was the one with the alpha voice. The warm wisdom of time and kindness is blasted with a barking testament to that person's know-how. The person willing to burn down the world to get the job done is the one around whom others seem to go silent. Something has gone terribly wrong with a culture that does not defer to calm consideration of context.

By contrast, I mostly see the room defer to the least sympathetic, most myopic, totally decontextualized loudmouth at the table. Somehow deep and wider wisdom of context is never on the agenda. The council of wisdom is not invited into the five-year plan. Next quarter's profits are bereft of the nth-order effects of any mono-cropped decision.

The actions that are ratified in those rooms spin into destructive consequences that destroy more and more and more. This is possible when there is selective "care." Care for the profit but not the people, care for the people but not the oceans, care for the rich but not the poor, care for the businesses but not the schools....

While it may appear that the alpha barking reductionist is the strongest person in the room, it is not true. They are the second strongest person. There is one person who is stronger than the strongman (who could be any gender). And that is the person who sees context and speaks from affection for life itself. This is not a pick-and-choose position. This is not a left political voice, nor is it a right political voice. On the contrary, this is the voice that reaches beyond those Venn diagrams and into macro and simultaneously micro attention to interrelations that produce life. Most boardrooms are desperately missing this voice.

I think most people, in some way, are trail angels already. Perhaps not explicitly, but in invisible practice, each of us is acting into the minutiae of doing that is wrapped into generations of tending. The counter gets wiped, the socks get washed, the baby is changed, and through these small acts, the world is massaged a little. The effort is dissipated, diffused, and written into every moment of the day.

The things we do without noticing we are doing them ... those things are the things children are learning the most from. They repeat those behaviors that their parents never once mentioned, such as how to be in an argument, who gets to sit where at the table, and who gets up from the table to get water for everyone when it has been forgotten.

There are thousands of transmissions made each day in the household that communicate the limits of cohabitation, family hierarchies, and unspoken rules of respect. There is no need to broadcast these expectations; they are usually not even noticed until someone violates the unwritten code.

Every family has implicit scaffoldings around their interactions throughout the day, but especially around food, the table, and the rituals of dining. It can be surprising to spend time with a family that has a different code than the one that is so familiar.

Unpredictable responses surprise the system; the conversation is wild, perhaps it may feel unsafe. I remember eating at a friend's house when I was a child and watching closely, where to sit, who spoke, and what sort of things they talked about.Eating with other families requires that one be at least a little bit of an anthropologist.

Watch quietly, wait for cues.

Are they going to say grace?

Are they reaching or asking for things to be passed across the table?

I remember once joining a family that signaled the need to pass the dishes of food with silent glances. They did not even ask; they depended upon each person noticing who did not have a serving of this or that. By contrast, other families just reach across one another and grab the butter or salt, and that gesture is a jovial, raw display of affection.

Some people speak loudly and yuck-it-up at the table, others eat in silence. Both are declarations of respect. There are infinite possible displays of honoring one another at a meal, and all of them can be twisted into their opposites with only a little bit of bitterness. The words, "Please pass the pepper," can be said in ways that mean, I love you, and in ways that say, I despise you.

These are the invisible cultures of the table that extend into life in all sorts of unexpected ways. There is a sort of spinal logic that imprints on how it is possible to be in the world in many contexts.

The word "care" brings both a notion of tending with love and the baggage of manipulation, humiliation, patronizing, and moralizing.

How many painful moments have been justified in the words, "I am only doing this because I care?"

The gift of doing something that is beyond care, beyond the altruism and benevolence, for someone, someday, somewhere, is an invitation to the nth-order of what might be a joyous appetite, a nourishment not listed or measured on the labels at the grocery store. Not so much is going according to the linear outcomes anyway.

Nurturing the unknown is not so far-fetched.

Fig. 50. Rachel Hentsch. (2026). [watercolor]

Fig. 51. Noah Qwarfordt. (2025) [acrylic]

ecology of change

The restlessness of sorrow makes each room a place to leave.

Isn't it stagnant in here? Isn't it late already? Time to be
in motion?

Does anyone else smell smoke?

Jumpy to go, to shift, to twitch to slide to scratch — the
kitchen is piled with nothing to eat — the bedroom is
pillowed with nothing like sleep,

My body is fidgety to do something where there is only
what cannot be undone — making long phone calls in the
time set aside to be alone — walking, running, pacing,
circling, writhing — slipping out of the grip of möbius
memories,

Stillness aches unquietly ...

The scaffolding of my world is being rebuilt,

The banging is annoying — there is nowhere left to rest in
the old versions of the story and the new version has yet to
take form.

All the furniture is gone.

continued ...

The house of my familiar has morphed.

My heart is sonar pinging off all of the new walls

Finding the doorways and windows

Where is my outline?

Where is my underline, my intertwine, my storyline?

Watery, fluid, salty amoebic beginnings are also endings
flushed with tears.

Blustery wind carries the fluff of the dandelions

The tall grass moves like the sea,

Spring for all its newness is ancient.

Old pain, new learning, old learning new pain.

This is the amniotic fluid of an ecology of change.

Fig. 52. Leslie Thulin Eubanks. (2025). [watercolor]

earned innocence

Picasso, who had a dark side I won't go into here, once said, "It takes a long time to become young."

The fact of his cruelty is not insignificant here. Nor is it innocent.

I do not see innocence as weakness or purity, but rather as something that comes occasionally with depth of experience, beyond the logic of living in cynicism. Naïveté, by contrast, I would say is perhaps more pure, perhaps more vulnerable; unbroken and inexperienced. But the naivete of pre-experience is not innocent; innocence, for me, is not easy to come by.

Perhaps we, as human beings, don't arrive into life innocent. Perhaps innocence is something that has to be earned. Truly, there is every reason to live in distrust, despair, suspicion, and vengefulness — these are the rotting carrion of cultures of greed. This is the putrified inheritance of centuries of harm. Since I am not a vulture who can readily eat this mess, I have to boil it, burn it; I have to go through the fire first.

In the brewing, stewing story of my experiences of being alive so far, I have only begun to find a few crumbs of innocence. But I feel them accumulating. There is no shortcut to the sort of innocence I am speaking of. It is a lightness that is found when all the fat of expectation has been roasted off.

When there is nothing left, there is suddenly everything. Grapes become wine, and then a red wine undergoes reduction to a sweet glaze, or butter clarified into ghee, or milk fermented into cheese. Time is an intrinsic ingredient of transformation altering chemical make-up into a different version altogether. But still somehow related. I am me, as I was when I was little, I have been through the whirling blades of life, and as I enter elder years I am finally becoming more of that essence I was as a child, not naively, but through exertion and rolling in the waters of life like a stone becoming a river rock, my softness is forming through strength. I am being digested by life.

The rhythm is to weep and scream and break over and over again, and in the interims sip sweetly from the nectar of this earth's offerings. In my life, I have seen the who and what of assumptions of the real, true and good, explode, crash, and shatter. This is the necessary arc.

My heart has broken, my identity has fractured, my thoughts have cracked. I have trusted people who I should not have. I have hurt people I did not mean to. The cells of my body hold stories of violation, the sinews of my heart know betrayal. I have carried the weight of my own spirit when it turned to stone. And when my own pain is not enough, witnessing my children's is the next wave to topple me. For most people, most of life is unthinkably vile.

But there is also breath, ecstasy, and falling in love. There is music, humor, and good food. There is the way we have all always been among the entire living world, and not in isolation. There is movement.

The triage is not to reassemble the broken heart or the broken world but to digest the rot that would be a belly ache into a belly laugh. Not to bypass, but to reclaim the morale to pick up and start again. Often through hardship, and each time with a little more attunement to the hiding poltergeists. I most likely cannot banish these ghouls since they live in me, but I can log their lairs. I want to be in the shifting of the matrix. I do not want to carry these damnable phantoms anymore.

After smashing headfirst into schismogenic situations long enough I am beginning to seek more alive approaches. Art is digestion of perception. Like water, art refracts perception. The nth-order of the artist's perception as it transfers into other forms is like saliva and bile that allow an apple to enter the mouth and metabolize into a seemingly non-correlated action. A laugh. A cry. A walk.

The artist is a cultural digestive agent, turning banality into the sacred, and the sacred into banality. Some art bleeds, and makes love out of anger. Some art is softer. Some art is manipulation. The offering is an indirect description, the love song, the use of light in the painting, the gestures in the sculpture, the framing of the photograph.

These are not only skills of the artists, but are the expressions of an arduous process of moving perception through the labyrinth of chewing, swallowing, soaking in acid, folding over and over in membranes of communing, shitting, soiling — until it arrives anew. An orange becomes a testament to sex, a few chords of a song evoke stories untold, a carved stone becomes a familiar tragedy.

Emily Dickinson said "tell all the truth but tell it slant." She is making room for movement. Art lets us say a thing, and then unsay it, and then resay it until the saying is a forest floor into which a world can seed. Artists are stomachs that receive one sort of perception, and offer it back translated into unknown languages. A culture is alive when it has the hiccups.

My mother and I brought great pain to one another, as so many parents and children do. When I was caring for her in her 90s, I was also preparing myself to store the memories of our strife when she had gone beyond. I knew the time would come after her death when those memories would become my stomach acid. Ingurgitating and regurgitating them into the future did not appeal to me, but I did not really know what else to do. I could intellectually forgive and forget, but the cells remember. There was no reason to imagine otherwise. It has not been easy to re-assimilate our relationship into a story to nourish rather than rancidify. Nourishment is like innocence in this context; it requires a metabolic heat to change its alchemy.

The guttural wonder of the belly laugh is not an overlooking or an undercutting of the horrors of the world, it is what is possible when that horror has soaked in the caustic fluid of digestion, emulsifying the story; fusing horror into humor and care, into delight in life. It takes time to tell it slant.

A shallow delight in life is nice, but it does not carry the ages like the delight that has risen through the festering pain of loss.

The microbiome is becoming more understood in its assistance of digesting food, but the bacteria that exist in our bodies are symbionts for the formation of physiological processes. We are of the bacterial realm. The digestion of experience into new perception requires making sense anew. Little did most of us realize that our sense organs were made by bacteria that needed stimulation. The experiences in a life are metabolized into ideas, language, habits, and next generations' trajectories. (Margulis, n.d.) The entire biosphere is ingesting and transforming everything into something to be ingested and transformed, formed, informed, reformed, deformed and efformed.

The whole arc takes water, it takes something like enzymes. But water and enzymes accompany histories of conversations between every organism in every size, shape and species. The earned innocence arrives with loss of carriage for former perception. I no longer can hold my mother in contempt. The cup in which to hold that contempt is no more, and therefore the contempt had nowhere to pool. Loss of carriage is distortion of the past, contortion of the communication, and absorption of learning.

I knew I wanted to be at peace with my mother, but peace as a concept was abstract until it moved through the transforms of my world, my body, and finally into and below words ... I had my ideas of what peace would look like, but they were linear, trite, and out of rhythm with the nth-order of healing. My abstract version of peace had little to do with the many contexts of situation that would untangle our eventual friendship. I could not have drawn a line or articulated what that could be, and if I had, it would have been scripted into a tight formula of what a good mother-daughter relationship looks like according to current family therapy models (whatever they may be). My visioning of what our relationship could be did not and could not possibly meet the profound possibility awaiting in nth-order.

What does it mean to contribute to a possible world in which distrust is not the necessary consequence of the surrounding logic?

How might the refraction of my perception of the world, or yours, become an angle into innocence?

Certainly, the fragmentation of being alive in a world of reductionism has forced strange sense-making for survival. To synthesize the needs of being an interdependent organism in a social system of isolated independence is contorting the vitality of necessary interrelationship ... people need to need each other. Needs get twisted up.

It will require many forms of nourishment to abate zeal for success in a destructive system.

There is no promise that medication or meditation, or psychedelics will give this, nor will one god or another (all the religions, though they offer necessary havens, are fraught with issues) — academia and knowledge is not an antidote, nor is health, or self-help workshops, or journaling. All of these things could help, or they could become more of the same problem, replacing one cove of solutions for another, none of which gets out of the pattern.

Sometimes being in nature and exposure to other cultures can help; sometimes a harsh heartbreak, an illness ... but there is no sure bet. Many people have been through all of these supposed cures and come out just as tangled as they were going in. Even art won't save us. Go home.

In the morning, I make my way to the kitchen. First, my bare feet touch the ground. I find my balance, and move through time and space to the faucet. I try to recall the movement of the water that fills the kettle. The rain, the clouds, the cycles of moisture through the trees and soil. I remember oxygen. I breathe. I am half asleep, and I know very little about plumbing, but I am grateful it works today.

I remember my mother standing at the sink in the morning banging the dishes as she made her tea. I still believe she was trying to wake us all up with all that clanking and noise. She claimed she was trying to be quiet. She was very deaf without her hearing aids in the morning, so I can't be sure.

I am in the minutiae of the fabric of the entire socio-economic cultural history in this tiny moment of making tea. I remember the Japanese tea ceremony. I bring my beloved a steaming cup into our barely lit room. The creature that is my body shapes into his contours. His belly is warm. The day will soon come rushing in, he smells like sleep.

The cup of water that gives itself to thirst is a model for me of the compassion that gives itself freely. Water is generous, tolerant, does not hold itself apart, lets itself be used by any need. Water goes, as Lao Tzu says, to the lowest places, vile places, accepts contamination, accepts foulness, and yet comes through again always as itself, pure, cleansed, and cleansing. (Le Guin, 2016)

references and contexts

Abraham, R. (2006). *Chaos, Gaia, Eros: A chaos pioneer uncovers the three great streams of history.* HarperOne.

Bateson, G. (1972). *Steps to an ecology of mind: Collected essays in anthropology, psychiatry, evolution, and epistemology.* Chandler Publishing Company.

Bateson, G. (1975). *Ecology of mind: UCSC seminar No.* 8 [Seminar]. University of California, Santa Cruz.

Bateson, G. (1979a). *Mind and nature: A necessary unity.* E. P. Dutton.

Bateson, G. (1979b). Letter to the University of California Regents. In *The Esalen catalogue.* Esalen Institute.

Bateson, N. (2016). *Small arcs of larger circles: Framing through other patterns.* Triarchy Press.

Bateson, N. (2023). *Combining.* Triarchy Press.

Borges, J. L. (2007). *Labyrinths: Selected stories and other writings.* New Directions.

Boye, K. (2014). Yes, of course it hurts when buds are breaking (D. McDuff, Trans.). In *Complete poems of Karin Boye.* Bloodaxe Books. (Original work published 1935)

Carson, R. (1962). *Silent spring.* Houghton Mifflin.

Guddemi, P. (Ed.). (2026). Ecology of mind — The course. Triarchy Press.

Hoffmeyer, J. (2008). *Biosemiotics: An examination into the signs of life and the life of signs.* University of Scranton Press.

Korzybski, A. (1933). *Science and sanity: An introduction to non-Aristotelian systems and general semantics.* Institute of General Semantics.

Le Guin, U. K. (2016, November 8). *The election, Lao Tzu, a cup of water.* https://www.ursulakleguin.com/blog/119-the-election-lao-tzu-a-cup-of-water

Lipset, D. (1982). *Gregory Bateson: The legacy of a scientist.* Prentice-Hall.

Margulis, L. (n.d.). *Lynn Margulis lecture* [Video]. YouTube. https://www.youtube.com/watch?v=863SYiQPqPo

Online Etymology Dictionary. (n.d.). Quicken. I*n Etymonline.* https://www.etymonline.com/word/quicken

Peirce, C. S. (1931–1958). *Collected papers of Charles Sanders Peirce* (C. Hartshorne, P. Weiss, & A. W. Burks, Eds.; Vols. 1–8). Harvard University Press.

Rappaport, R. A. (1974, Summer). Sanctity and adaptation. *The Co- Evolution Quarterly*, (12), 59.

Stevens, W. (1952). *The man with the blue guitar, including ideas of order* (2nd ed.). Alfred A. Knopf. (Original work published 1937)

art credits

table of contents (alphabetical order)

Mother and daughter in the kitchen.

Our microbes seasoning all that we touch, as time and ecologies stew.

Not pictured here are the many mothers that came before.

Granddaughters, and grandmothers.

We feed each other.

We learn together.

Lois and Nora Bateson.

Photograph by Barry Schwartz, © 2001.